Library of
Davidson College

Evolutionary Changes
to the
Primate Skull
and
Dentition

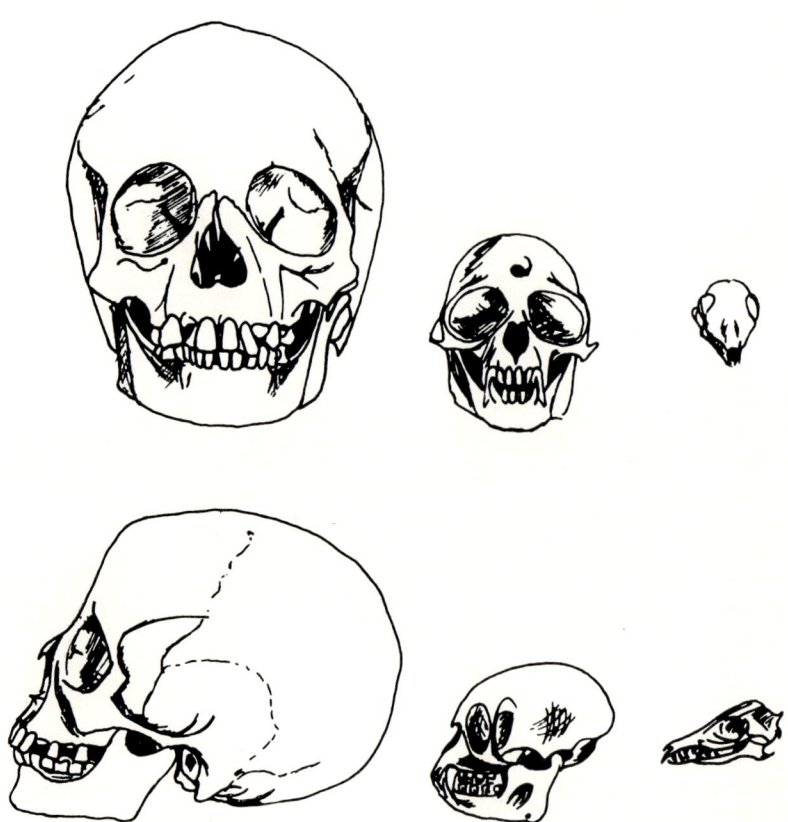

Evolutionary Changes
to the
Primate Skull
and
Dentition

By

C. L. B. LAVELLE, D.Sc., Ph.D., M.D.S.

*Head of the Department of Oral Biology
Faculty of Dentistry
The University of Manitoba
Winnipeg, Canada*

**R. P. SHELLIS, Ph.D.
D. F. G. POOLE, Ph.D.**

*MRC Dental Unit
Dental School
University of Bristol
Bristol, England*

CHARLES C THOMAS · PUBLISHER
Springfield · Illinois · U.S.A.

Published and Distributed Throughout the World by
CHARLES C THOMAS • PUBLISHER
BANNERSTONE HOUSE
301-327 East Lawrence Avenue, Springfield, Illinois, U.S.A.

This book is protected by copyright. No part of it may be reproduced in any manner without written permission from the publisher.

© 1977, *by* CHARLES C THOMAS • PUBLISHER
ISBN 0-398-03618-7
Library of Congress Catalog Card Number 76-41880

With THOMAS BOOKS *careful attention is given to all details of manufacturing and design. It is the Publisher's desire to present books that are satisfactory as to their physical qualities and artistic possibilities and appropriate for their particular use.* THOMAS BOOKS *will be true to those laws of quality that assure a good name and good will.*

Library of Congress Cataloging in Publication Data

Lavelle, Christopher Lawrence Bannerman.
   Evolutionary changes to the primate skull and dentition.

   Includes bibliographies and index.
    1. Primates—Evolution. 2. Skull. 3. Teeth.
I. Poole, David Frederick George, joint author.
II. Shellis, R. P., joint author.   III. Title.
[DNLM: 1. Primates. 2. Evolution. 3. Skull.
4. Dentition.   GN281.4 L399e]
QL737.P9L33   599'.8'0438   76-41880
ISBN 0-398-03618-7

Printed in the United States of America

to all those who have interests in the skull, jaws, and teeth: they have a great responsibility in the elucidation of primate evolutionary changes

INTRODUCTION

RECENT YEARS have witnessed a significant improvement in the condition of primate, especially hominoid, nomenclature. This improvement reflects three factors. First, every new fossil has ceased to be regarded by palaeontologists as a new species. Secondly, palaeontologists, zoologists and physical anthropologists have expanded the concept of variability in palaeospecies. Hence, greater ranges of morphological variation are lumped together within a single species. Finally, the International Code of Zoological Nomenclature has been adopted by the majority of workers.

Physical anthropology occupies a pre-eminent niche in biological research. This is not only a consequence of its intrinsic intellectual interest. Indeed, only by critical investigation of extant and fossil primates is it possible to deduce functional and evolutionary data of particular morphological structures (or features). Despite the wealth of literature published each year, however, many fundamental questions have yet to be evaluated. A considerable proportion of the current anthropological literature remains repetitive and contributes little to the understanding of the fundamental principles encompassing primate morphology.

In spite of active and often preferential researches, palaeontological knowledge of primates remains fragmentary. Since primates are fundamentally linked to tropical forests—an unfavorable environment for fossilization—many gaps may never be filled. Nevertheless, the most ancient radiations, and therefore the first radiations, are already better known. We are, therefore, leaving behind the confusion that has enveloped the history of the Hominoidea. Among extant primates, zoologists and biochemists generally agree to recognizing four natural groups, often considered as infraorders: *Lemuriformes, Lorisifomes, Tarsiiformes,* and *Simiiformes* (Anthropoidea). All post-Eocene fossils, and most of those from the Eocene, can be assigned to one or other of these taxa. It seems, therefore, that the four extant groups differentiated during the early Tertiary. The possible exception is the *Lorisiformes*, which are known since the Miocene only. The more ancient fossils, especially those from the Paleocene, to which must be added their direct descendants in the Eocene, present, together with evidently primitive features, some unexpected specializations in such ancient forms; they represent an early radiation and do not belong to the recent infraorders.

Hominoids—chimpanzee, gorilla, and man—have apparently all evolved during the past five to fifteen million years from a common ancestor, *Dryopithecus*. The field of human evolution is particularly concerned with the particular ecological features which have selected for behavioral and structural traits found in the divergent hominoid line which evolved into *Homo*. Competition in a common environment among sympatric populations of the ancestral species of the Pongidae and Hominidae, along with competition from other primates and carnivores, may possibly have induced a divergence which led to varied econiches for hominids and apes. The geographical econiches distribution of contemporary primates seems to indicate that competitive exclusion has been operating extensively. Gorillas, for instance, live in mountainous regions and are largely vegetarian. Chimpanzees are suited to a terrestrial environment, although they have retreated into forest environments in every region where they have been sympatric with modern man. Baboons are also terrestrial, plain-living animals and, like humans, are carnivorous hunters. Moreover, humans and baboons may have seriously competed for similar food sources at times in their phylogenetic histories. A slightly unequal reliance on different food sources and incipient tool use may, however, have enabled hominid ancestors a competitive advantage while other terrestrial apes came to rely more on woodland-forest environments. Conversely, the pongid ancestors might have competed more successfully in the forest-woodlands and edged the protohominids onto the plains. Once the populations begin exploiting different econiches, however, the resulting allopatry would promote further divergence by limiting gene flow.

Whether dietary specialization or incipient tool use came first is debatable, as is the primary reason for the initial hominid divergence. There is little doubt that once econiche diversity for Miocene apes was effected, the different econiches provided varying opportunities and selective pressures for culture and that culture was instrumental at some point in influencing the divergence of protohominids and other hominids. How this is reflected in the morphology of the primate skull is the subject of this book.

Morphological investigation of extant primates remains the kernel of investigations into fossil specimens. Such investigations are generally limited to techniques of dissection of soft tissues and observations of the hard underlying structures coupled with the association of these to functional behavioral developments. The principal method of studying form still relies on the experienced eye and the creative mind behind the eye. Additional information is often acquired by univariate or bivariate analysis of simple measurements. Combinations of such measurements, in the guise of indices, are frequently computed for the analysis of shape. Such elementary metrical study rarely achieves more than a mere confirmation of data already obtained from subjective visual analysis. Indeed, for accurate objective assessment of skeletal form, many dimensions must be measured coupled with their subsequent analysis by multivariate statistical techniques. Only in this way can differing modes of variation and multivariation coupled with varying kinds of correlation that characterize the metrical definition of complex shape be accommodated.

Multivariate statistical analysis is capable of allowing for such perturbations of data that are difficult to evaluate visually and impossible to reveal visually or by simple measurement alone. There are, however, some restrictions or limitations to the wholesale application of multivariate statistical analysis in morphological studies.

First, measurements only provide data in relation to datum points. They do not provide information about shape between the datum points. Obviously, therefore, an increase in the number of datum points will improve the metrical definition of shape, although there may be practical problems in recording a large number of dimensions.

Second, measurements frequently depend upon the particular orientation of specimens along standard lines or planes. In actual fact, there is no a priori reason why homologous lines or planes may not be curved, with their curves varying in different species. This facet is difficult to accomodate unless both sophisticated measurement and statistical analytical techniques are employed.

Finally, the choice of datum points is always a subject which arouses controversy, since quite frequently the selection is based on subjective intuition. It is obvious, however, that no matter how sophisicated the method of statistical analysis, definitive conclusions depend ultimately on the existence of sound data.

When all the characters are measured, measures of statistical distances between populations are little more than measures of size differences. Distance is, of course, not purely a measure of size difference, since shape and morphology are defined by differential size. But differences of magnitude greatly outweigh the effect of differences of proportion in distances whenever the taxonomic units vary more than a slight amount in size.

Besides the taxonomic shortcomings of pure size differences, they may be further rendered inaccurate or statistically invalid by failure of the raw data to conform to the assumption of distribution normality, large sample size, equality of variance, and linearity and homogeneity of covariance upon which the probability theory depends. The conformity of input data to these conditions is not commonly documented or defended by researchers before the use of statistical routines such as an analysis of variance, factor analysis, and canonical variate analysis. Far more analysis and checking of the original data is therefore required before elegant and sophisticated statistical techniques can be applied.

In addition, there is also the concept of biological variation and the difficulty of obtaining representative samples. Indeed, when dealing with isolated fossil specimens, there is no chance of ascertaining whether the specimen is typical of a population or is representative of an extreme of the population range. Nevertheless, unless a metrical approach is adopted, there is no way that evolutionary studies can be elevated from the subjective to the objective plane.

The purpose of this book is to concentrate on the skull, since compared with the remainder of the skeleton, the cranium, especially the teeth, is preferentially preserved. Also, while considering the skull of primates as a whole, attempts are made to concentrate on hominoids. This results from the fact that hominoid skulls comprise a number of morphological features which have remained controversial for far too long, and it is high time that a more constructive approach was adopted in order to eradicate the majority of these subjective and controversial assessments. Nevertheless, until more hominoid skulls are described and the hominoid evolutionary lineage fully interpreted, an objective interpretation of the important morphological features will remain incomplete. Furthermore, until nonhuman primate skulls received the detailed investigation afforded to human skulls (e.g. Downs, 1956; Bjork, 1960), little objective data will ensue. This book is divided into a number of sections relating to the neurocranium, the facial skeleton, the jaws, and the teeth. Attempts are made, however, to emphasize that the skull is a complex biological unit as a whole rather than a series of discrete biological structures. Far too often, evolutionary changes in one morphological attribute are interpreted without due regard to the biological significance to the skull as a whole.

Relative brain size is defined as the ratio of the actual to the expected brain size. The expected brain size may be computed from the regression equation in which brain size is predicted from body size. Jerison (1970) has analyzed the history of brain size in Teritiary animals and their living descendents. He has shown a continuing increase in relative brain size with carnivores having relatively larger brains than their ungulate contemporaries. This brain size increase has also been shown to parallel the evolution of greater diversity. But despite the evident general trend towards an increase in average brain size, there is an interesting and important overlap in the region of low brain size. This indicates the presence of at least some small-brained species at all times. Thus, the evolution of enlarged brains, though generally a route to success and survival of new species, was apparently not universal even among progressive orders. The key factor is probably that the brain of mammals has evolved in ways appropriate to behavior within particular niches. As more diversified niches were invaded, more diversified brain adaptations evolved. Nevertheless, the fact that some animals continued with a small brain size suggests that the latter does not lead to extinction in some groups.

It is also interesting that, from examination of asymmetry in mountain gorilla skulls (*Gorilla gorilla beringei*), Groves and Walker (1973) concluded that cerebral form and function have little effect on skull shape.

A constant supply of hominoid fossil specimens is being described by Leakey and his coworkers (e.g. Leakey and Wood, 1974), and these may well shed some important light on the pattern of human evolution. There are, however, one or two problems. First, there is an overwhelming need for some growth data on fossil specimens. Even when considering modern man, there is little accurate longitudinal data (Knott, 1972). The second problem is the degree of association between one part of the skull and another. For instance, during growth, associations have been described between the upper and lower jaws (Slavsgold, 1971), the mandible and cranial base (Hoyte, 1971; Droel and Isaacson, 1972; Knott, 1973), and between the nasal septum and jaws (Sarnat, 19700). Thus, there are varying degrees of associations between one part of the skull and another (Solow, 1966; Brown, 1969; Mitani, 1973), although whether there are similar patterns of association in nonhuman primate skulls has yet to be elucidated. In addition to metrical traits, Hertzog (1968) has listed some associations between discontinuous cranial traits.

Although there is considerable data relating to the evolution of human skull form (Bunak, 1968) and many detailed metrical analyses of the human skull (Stoessiger and Morant, 1932; Little, 1943; Tattersall, 1968), the relationship between genetic and environmental factors to skull form has received little study (McKeown, 1974). Nevertheless, from the existence of secular changes in skull form (Barnard, 1935; Ingervall, Lewin and

Hedegard, 1972), it is evident that environmental factors do play a role.

Possibly the most fruitful method of analyzing the degree of variation in skull dimensions involves factor analysis. For instance, from fifty-four measurements of one hundred Anglo-Saxon skulls, Howells (1957) selected three measurements of length, breadth, and height which accounted for 51 percent of the variation in all measurements with the matrix of residual correlations yielding a further seven factors representing regions of variation not related to the influence of the first three. These seven additional dimensions accounted for a further 31 percent of the total variance and included variation in supraorbital development, in forebrain width, in forehead fullness, in fullness across the top of the parietals, in lower occipital fullness and in basal breadth. These ten measurements, therefore, accounted for virtually all of the correlation in the cranial vault proper and each were virtually independent of one another.

In a later study of Egyptian crania, Landauer (1962) extracted five factors from the facial skeleton, one of general size of the skull as a whole, one of the size of skeletal mass and muscular strength, one of the breadth across the malar bones, one of frontal fullness, and one of facial breadth. In another study based on cranial angles and indices, Chopra (1969) compared twenty different cranial series using factor analysis. On the average, five factors covered the total variation, although the factor structure differed considerably from group to group, so that seventeen different factors were identified. These included cranial and facial width, facial relief, cranial ruggedness and the degree of prognathism.

If only a similar study were performed on nonhuman primate skulls, then there might be a considerable advance in the knowledge of primate evolutionary trends.

Dental arches (or dental arcades) have occupied limited status in the study of primate evolution. Possibly one feature responsible for this lack of status stems from the dearth of longitudinal data on the growth changes in the dental arch (Knott, 1972). Nevertheless, an example of the value of the dental arch in primate taxonomy can be illustrated from *Ramapithecus*. Species of *Ramapithecus* are among the few hominoid species currently considered as possibly close to the direct line of human ancestry, with one of the most striking resemblances between *Ramapithecus* and later hominoids being the supposed presence of parabolic dental arches. In a recent study, Walker and Andrews (1973) have reconstructed the dental arcades of *Ramapithecus wickeri*. These indicated very small incisors and posteriorly diverging and nearly straight cheek tooth rows rather widely separated and curved tooth rows such as are found in modern man. The palate is narrow and the tooth rows are relatively very elongated. The premaxillary regions are most abbreviated and a small snout with a small piriform aperture projected from a broad and very flat face which was especially wide in the zygomatic region. These peculiar and unique gnathic features clearly place *Ramapithecus wickeri* apart from contemporary species of *Dryopithecus*.

There is, however, still far too little quantitative data relating to the nonhuman primate dental arch, in particular how dental arch form is related to that of the skull.

Recently there has been a search for factors to discriminate the hard palates between human and nonhuman primates. The size of the foramen incisivum is just one supposedly discriminating feature. From a study of Eskimo and Bavarian skulls, however, Helmuth (1973) has shown that this foramen is partly associated with the size of the palate. This, therefore, emphasizes another feature, namely that size factors must be eliminated as far as possible in order to obtain critical primate taxonomic classifications.

The role of genetic and environmental factors in tooth development has long been controversial (Chung, Niswander and Runch, 1971; Potter, 1972; Anderson and Thompson, 1973; Wickramaratne, 1974; Johanson, 1974; Portin and Alvesalo, 1974). Yet definitive evidence on their respective roles is essential in order to account for the reduction in tooth size during hominid evolution (Sofaer, 1974). Even with the sequence of tooth eruption, there is little data concerning the role of genetic and environmental factors (Garn, Wertheimer, Sandusky and McCann, 1972;

Shumaker, 1974). Nevertheless, in view of the existence of secular changes in tooth size (Eberling, Ingervall and Hedegard, 1973), environmental factors exert some role in dental development, but is it possible that dental and skull development share similar genetic influences?

Tooth attrition has been examined from an experimental (Brace and Molnar, 1967) and cultural (Molnar, 1972) viewpoints. Nevertheless, as tooth attrition is a universal primate phenomenon, it is difficult to ascertain its taxonomic significance. Looking at attrition from another aspect, one of the distinguishing features between robust and gracile australopithecines is traditionally the different chipping of the teeth. From examination of worn deciduous and permanent teeth, Wallace (1973) concluded that there was no significant difference between the chipping of these two forms. This worker therefore concluded that there was no difference in the diet of robust and gracile australopithecines. However, the temporomandibular articulations, the insertions of the muscles of mastication, and tooth size have probably more significance in identifying diets compared with tooth chipping.

From these unanswered questions, it is obvious that despite its illustrious study, there is still little definitive evidence on the evolutionary changes of the primate skull. At the end of this book, therefore, it is hoped that the reader will be convinced of a new direction for his research, particularly in view of the recent fossil finds emanating from Africa (Oxnard, 1975).

REFERENCES

Anderson, D. L. and Thompson, G. W.: Interrelationships and sex differences of dental and skeletal measurements. *J Dent Res*, 52:431, 1973.

Barnard, M. M.: The secular variations of skull characters in four series of Egyptian skulls. *Ann Eugen*, 6:352, 1935.

Bjork, A.: The relationship of the jaws to the cranium. In Lundstrom, A. (Ed.): *Introduction to Orthodontics*. London, McGraw, 1960, pp. 104-140.

Brace, C. L. and Molnar, S.: Experimental studies in human tooth wear. *Am J Phys Anthropol*, 27:213, 1967.

Brown, T.: Facial growth pattern and coordination. *Aust Orthod J*, 2:5, 1969.

Bunak, V. V.: On the evolution of the shape of the human skull. *Vopr Antrop*, 30:3, 1968.

Chopra, V. P.: Faktoren des Craniums bei prahistorischen Populationen. *Homo, 20*:185, 1969.

Chung, C. S., Niswander, J. D., and Runck, D. W.: Genetic and epidemiologic studies of oral characteristics in Hawaii's school children: dental anomalies. *Am J Phys Anthropol, 36*:427, 1971.

Downs, W. B.: Analysis of the dentofacial profile. *Angle Orthod, 26*:191, 1956.

Droel, R., and Isaacson, R. J.: Some relationships between the glenoid fossa position and various skeletal discrepancies. *Am J Orthod, 61*:64, 1972.

Ebeling, C. F., Ingervall, B., and Hedegard, B.: Secular changes in tooth size in Swedish men. *Acta Odont Scand, 31*:141, 1973.

Garn, S. M., Wertheimer, F., Sandusky, S. T., and McCann, M. B.: Advanced tooth emergence in negro individuals. *J Dent Res, 51*:1506, 1972.

Groves, C. P. and Walker, N. K.: Asymmetry in gorilla skulls: evidence of lateralysed brain function. *Nature, 244*:53, 1973.

Helmuth, H.: On the foramen incisivum in *Homo sapiens*: Morphology and phylogenetic aspects. *Homo, 24*:117, 1973.

Hertzog, K. P.: Associations between discontinuous cranial traits. *Am J Phys Anthropol, 29*:397, 1968.

Howells, W. W.: The cranial vault: factors of size and shape. *Am J Phys Anthropol, 15*:19, 1957.

Hoyte, D. A.: Mechanisms of growth in the cranial vault and base. *J Dent Res, 50*:1447, 1971.

Ingervall, B., Lewin, T., and Hedegard, B.: Secular changes in the morphology of the skull in swedish men. *Acta Odont Scand, 30*:539, 1972.

Jerison, H. J.: Brain evolution: new light on old principles. *Science, 170*:1224, 1970.

Johanson, D. C.: Some metric aspects of the permanent and deciduous dentition of the pygmy chimpanzee (*Pan paniscus*). *Am J Phys Anthropol, 41*:39, 1974.

Knott, V. B.: Longitudinal study of dental arch widths at four stages of dentition. *Angle Orthod, 42*:387, 1972.

Knott, V. B.: Growth of the mandible relative to a cranial base line. *Angle Orthod, 43*:305, 1973.

Kurisu, K., Kato, S., and Kanda, S.: A factor analysis of the environmental correlation matrix from the craniofacial traits of male monozygotic twins in growing age. *Med J. Osaka Univ, 22*:207, 1972.

Landauer, C. A.: A factor analysis of the facial skeleton. *Hum Biol, 34*:239, 1962.

Leakey, R. E. F. and Wood, B. A.: New evidence of the genus *Homo* from East Rudolf. *Am J Phys Anthropol, 41*:237, 1974.

Little, K. L.: A study of a series of human skulls from Castle Hill, Scarborough. *Biometrika, 33*:25, 1943.

McKeown, M.: Genetic and environmental influence on the shape of the adult dog skull. *Angle Orthod, 44*:62, 1974.

Mitani, H.: Contribution of the posterior cranial base and mandibular condyles to facial depth and height during puberty. *Angle Orthod, 43*:337, 1973.

Molnar, S.: Tooth wear and culture: A survey of tooth functions among some prehistoric populations. *Curr Anthropol, 13*:511, 1972.

Oxnard, C. E.: The place of the australopithecines in human evolution. *Nature* (Lond), *258*:389, 1975.

Portin, P. and Alvesalo, L.: The inheritance of shovel shape in maxillary central incisors. *Am J Phys Anthropol, 41*:59, 1974.

Potter, R. H.: Univariate versus multivariate differences in tooth size according to sex. *J Dent Res, 51*:716, 1972.

Sarnat, B. G.: The face and jaws after surgical experimentation with the septovomeral region in growing and adult rabbits. *Acta Otolaryngol, 268*:1, 1970.

Shumaker, D. B.: A comparison of chronologic age and physiologic age as predictors of tooth eruption. *Am J. Orthod, 66*:50, 1974.

Slavgsvold, O.: Associations in width dimensions of the upper and lower jaws. *Trans Eur Orthod Soc, 71*:465, 1971.

Sofaer, J.: Differential evolutionary reduction of tooth size as a consequence of skeletal reduction. *J Dent Res, 53*:752, 1974.

Solow, B.: The pattern of craniofacial associations. *Acta Odont Scand, 24*: suppl. 46, 1966.

Stoessiger, B. and Morant, G. M.: A study of the crania in the vaulted ambulatory of St. Leonard's Church, Hythe. *Biometrika, 24*:135, 1932.

Tattersall, I.: Multivariate analysis of some medieval British cranial series. *Man, 3*:284, 1968.

Walker, A. and Andrews, P.: Reconstruction of the dental arcade of *Ramapithecus wickeri*. *Nature, 244*:313, 1973.

Wallace, J. A.: Tooth chipping in the australopithecines. *Nature, 244*:117, 1973.

Wickramaratne, G. A.: The 'skeletal profile' and dentition of some inbred strains of mice. *J Anat, 117*:565, 1974.

ACKNOWLEDGMENTS

THE COMPILATION of this monograph could not have been attempted without the initial inspiration of Lord Zuckerman, while one of the authors (C.L.B.L.) was still an undergraduate student. This inspiration was subsequently rekindled by Professors E. H. Ashton, C. E. Oxnard, and P. M. Butler. Furthermore, very grateful thanks is indebted to Mrs. E. Okrainec, who has rescued me on more than one occasion. Finally, much appreciation is due to Dr. J. W. Neilson, Dean of the Faculty of Dentistry, the University of Manitoba, who encourages time and effort to be concentrated on research.

The work on Chapter Five could not have been carried out without the generosity and cooperation of many people who have donated the material for study, some of which was scarce. We are, therefore, pleased to express our gratitude to the following: Mr. P. Buck, Microbiological Research Establishment, Porton Down; Mr. D. Coles, Department of Dental Medicine, University of Bristol; Dr. K. S. Joysey, University Museum of Zoology, Cambridge; Mr. G. Lander, London Hospital Medical College; Dr. C. L. B. Lavelle, Department of Oral Biology, the University of Manitoba; Dr. J. Musgrave, Department of Anatomy, University of Bristol; Dr. P. Napier, British Museum (Natural History); Mr. M. Plant, Pharmaceutical Division, Fisons Ltd.; Dr. L. F. Taffs, New Polio Unit, National Institute for Medical Research.

We also thank Mrs. J. Barnett for skillful preparation of numerous ground sections; Mrs. G. Bennerson and Miss R. Porter for excellent photographic work; Mr. M. S. Gillett for assistance with scanning electron microscopy and Mrs. J. Powell and Mrs. S. B. Phillips for typing the manuscripts. Dr. J. E. Tyler kindly gave permission to publish Figures 24 and 29; Miss T. Ovenden, Department of Radiology, University of Bristol prepared many radiographs which were of great help in specimen preparation.

The authors are indebted to Pergamon Press for permission to reproduce Figure 25 and to the *British Dental Journal* for Figures 23, 28 and 32.

C.L.B.L.
D.F.G.B.
R.P.S.

CONTENTS

	Page
Introduction	vii
Acknowledgments	xix

Chapter

One:	Brain Size of Primates	3
	Introduction	3
	Synopsis of Evolutionary Changes to the Primate Brain	4
	Metrical Definition of Brain Size	8
	Endocranial Capacity and Culture	23
	Brain Size and Race	27
	Conclusion	30
	References	34
Two:	Primate Skull	40
	Introduction	40
	Synopsis of Primate Skull Form	42
	Size of Cranium	50
	Orbits	52
	Nasal Form	54
	The Balance of the Head	56
	Location of the Foramen Magnum	61
	Mastoid Process	62
	Quantitative Skull Analysis	62
	Cresting	65
	Statistical Analysis of Skull Dimensions	67
	Variation in Cranial Form	70
	Craniofacial Correlations	74
	Conclusion	78
	References	78

Chapter	Page
THREE: THE PRIMATE MASTICATORY SYSTEM	85
Introduction	85
Maxilla	90
Mandible	94
Occlusofacial Relationships	102
Temporomandibular Joint	103
Dental Arch	107
Conclusion	124
References	125
FOUR: THE PRIMATE DENTITION	130
Introduction	130
Control of Tooth Development	130
Qualitative Assessment of Tooth Shape	138
Odontometry and Quantitative Tooth Descriptions	150
Evolution of Hominid Tooth Form	159
Rates of Evolutionary Change in the Hominid Dentition Attrition	165
Attrition	170
Eruption Sequence	173
Variation in Tooth Number	175
Conclusion	184
References	185
FIVE: THE CALCIFIED DENTAL TISSUES OF PRIMATES	197
Introduction	197
The Dental Tissues of Man	198
The Dental Tissues of Nonhuman Primates	218
Discussion	260
References	277
Index	281

Evolutionary Changes
to the
Primate Skull
and
Dentition

CHAPTER ONE

BRAIN SIZE OF PRIMATES

INTRODUCTION

PALAEONTOLOGICAL STUDIES have shown that the Primates evolved from Insectivores (Campbell, 1969). Thenius (1969) stated that the initial attribute of the Primates may have been the development of prehensile hands and feet in relation to arboreality. Also, the beginning of arboreality was accompanied by enlargement of the eyes, a rostrally directed shift of the optical axis, shortening of the craniofacial skeleton and the development of a more voluminous endocranium. The enlargement of the brain is considered to have been an important, and to a certain extent specific, characteristic of primate evolution.

Until significantly more fossil data are available, the capacity of the endocranial cavity will remain controversial, particularly in relation to primates. Emotive nuances, especially when correlating brain size with intellectual capability, have overshadowed estimates of endocranial capacity. Indeed, although data have been based upon relatively few specimens, far-ranging conclusions have been deduced which cannot withstand critical analysis. This is remarkable in view of the innumerable research techniques which have been applied to the brain itself. Even today, possession of a large brain is still taken to be indicative of mental prowess and small brains with stupidity. Yet, if this concept is applied to *Homo sapiens,* only very small brains are inadequate, whereas very large brains seems to offer no advantage over medium-sized ones.

There are many examples of successful animals with small brains in the broader fields of biology. The traditional argument that dinosaurs became extinct as a consequence of their small brains is incompatible with the fact that these creatures were eminently successful and ruled the world for a hundred million years. Alligators and oppossums have small brains yet they appear successful, so that too much survival value cannot be attributed to the advantage of a large brain. Dinosaurs possibly became extinct due to their overspecialized morphology rather than their diminutive endocranial capacity.

Somewhere along one of the mammalian evolutionary lineages, man's ancestors began to develop a specialized part of the brain in response to the evolutionary advantage of adaptability, i.e. the antithesis of adapting to a special environment. In fact, the primate lineage exemplifies the expediency of not specializing morphologically but rather in exploiting an organ permitting rapid adaptation to new situations. This organ was the new part of the cerebral cortex (neocortex) comprizing principal connections with somatic sensory paths, tactile, auditory and visual, and this neocortex was added to the old olfactory cortex already well adapted in amphibians and reptiles. In amphibians, the neocortex is merely primordial, whereas in many reptiles it is represented by a recognizable area on the dorsolateral suface of the hemisphere. The mammalian neocortex seems to facilitate learning and in primates this cerebral enlargement has been dramatic. In addition to the enlargement of the neocortex per se, there has been an increase in its cortical associations with other regions. This development may be explained by the great survival value to the human species of an ability to communicate, and the association or interconnection between cortical and subcortical regions facilitates language development (Geschwind, 1964). Moreover, palaeontology suggests that a remarkable development of the hominoid brain occurred during the latter part of the Cenozoic era, and the human brain that then evolved seems to have achieved its present complexity and size by Neolithic times.

SYNOPSIS OF EVOLUTIONARY CHANGES TO THE PRIMATE BRAIN

Primates may be distinguished from other mammals not only by the tendency to develop large brains relative to total body size but also gradual cortical enlargement and elaboration. This latter is mainly associated with the reception of sensory impulses and their transformation into suitable behavior patterns. The primate evolutionary trend to enlarge and perfect the brain has also resulted in an increase in the ratio of brain weight to body weight compared with that of other mammals with comparable body size. There is also a general allometric relationship between body and brain size, whereby relative brain size reduces markedly with increased body size independent of other neural specializations. This accounts for the fact that a small, fully grown squirrel monkey possesses proportionately a larger brain compared with adult man, i.e. 1:30 as against 1:51.

The brain and neurocranium grow very rapidly during prenatal life, so that at birth the endocranial capacity has reached a much larger percentage of its final adult size compared with the body as a whole. Hence, relative brain size must be judged not only by considering the allometrically influential general body size, but also by the physiological age or stage of ontogenetic development. Only with cognizance of these factors is it possible to appreciate the advances which have occurred in the growth curves of the relative capacities of catarrhines compared with marmosets, in pongids compared with the former and in man over the condition in all three anthropoid apes. In absolute size, marked ranges and overlapping ranges of endocranial capacities occur so that undue conclusions based on fossil data are to be avoided, particularly when other morphological features are absent. Also marked asymmetries in size and in the configuration of the two hemispheres, preclude undue reliance based on the analysis of fossil primate brains (or their endocranial capacities). Nevertheless, it is instructive to glance at the evolutionary changes which have occurred to the primate brain, bearing in mind that when more data are available relating to both fossil and primate specimens, greater confidence will be placed on their interpretation.

The structure of the brain offers some of the most cogent information for the affinities of tree shrews with Primates, in addition to demonstrating their distinction from Insectivora—the former taxonomic category. For instance, the relative size of the brain as a whole, expansion of the neocortex and formation of a distinct temporal lobe coupled with the posterior projection of the occipital lobe provide contrasts between three shrews and insectivores. In contrast, the brain of *Tarsius* still remains more primitive than that of any other living primate as judged from the relationship between brain size as a whole and the relative size of the olfactory bulb. Unfortunately, there is no certainty about the size and proportions of the brain in the early fossil primates of the family Plesiadapidae, although the skull of *Plesiadapis* suggests a brain even more primitive than that of *Ptilocercus*.

Among extant lemurs, the most primitive brain is that of *Microcebus* in which the olfactory bulbs, although reduced compared with *Tupaia* and lower mammals, are generally better developed than in other lemuroid genera. The formation of a deep sylvian fissure is associated with a characteristic enlargement of the temporal lobe. The neocortex is more richly convoluted in larger lemurs and the olfactory lobe and olfactory area of the cortex relatively more reduced. In the larger lemurs, the sulcus centralis is commonly represented by no more than a cortical dimple between the motor and sensory cortical areas. In general, however, the absence of limiting sulci over the frontal and parietal regions and the presence of longitudinally disposed sulci are distinguishing features for the convolutional pattern of lemuroids from that of the Anthropoidae. Nevertheless, the potentiality for the development of a transverse pattern of limiting sulci in occasional species, e.g. *Perodictus*, points to the affinity between these two primate groups.

Endocranial casts of the Eocene *Adapis* indicate a brain which is more primitive than any lemuroid brain, in that it is smaller relative to body size, the olfactory bulbs are relatively larger and the cerebral hemispheres are abbreviated such that the whole of the cerebellum is exposed superiorly. Nevertheless, precocious enlargement of the temporal lobes characteristic of the evolution of the lemuroid brain is also evident in the endocranial casts of the early Miocene genus *Progalago*.

The brain of the modern *Tarsius* illustrates both advanced and primitive characters. In general appearance, the cerebral hemispheres resemble simian proportions, whereas the olfactory tract and cortex are correspondingly small. The cerebral hemispheres show no fissuration on their exposed surfaces, except for a short oblique sulcus which may be present in the sylvian fossa. The tarsoid brain is basically primitive upon which a very specialized one-sided development of the visual centers associated with a reduction of the olfactory center has been superimposed. The endocranial cast of the Eocene tarsoid, *Necrolemur,* indicates a brain almost identical with that of *Tarsius,* except that the olfactory bulbs are less reduced and the cerebral hemispheres do not so completely cover the cerebellum.

There is a marked advance in the cerebral development of the modern monkeys beyond the prosimian level, e.g. in marmosets, brain weight is approximately three times greater than *Tarsius* and *Galago* while sharing a similar body size. In *Gallithrix,* the cerebral hemispheres are voluminous, particularly in the frontal and occipital lobes with the latter being extended posteriorly to overlap the cerebellum and medulla almost completely. The olfactory lobes are also greatly reduced in size. The neocortex is smooth except for a deep sylvian fissure which is associated with marked development of the temporal lobes. The relative smoothness of the neocortex of the marmoset brain relative to that of most other primates is not related to the small body size but is presumably indicative of an essentially primitive characteristic.

In most catarrhines and platyrrhines, the cerebral cortex becomes richly convoluted especially in the region of the sulcus centralis and lunatus. This latter sulcus is so peculiar to monkeys and apes that it is often referred to as the "simian sulcus." On the dorsolateral surface of the hemispheres, the sulci are generally disposed transversely as limiting sulci: a patterned arrangement of sulci which provides a rather striking contrast with the common lemuroid type of pattern where the sulci tend to be longitudinally disposed.

Although simplified in form, the convolutional pattern in larger apes, especially *Gorilla,* closely resembles that of the human brain. The contrasts are basically associated with the greater expansion of the association areas in the human brain. One consequence of the parietal cortical expansion in *Homo sapiens* is that the visual cortex becomes displaced posteriorly and with this displacement there is almost complete disappearance of the "simian sulcus," although in some individuals the more primitive condition is retained including retention of the "simian sulcus." The convolutional pattern evident in monkey brains is basically similar to that of anthropoid apes, although this may be somewhat obscured by the secondary sulci. The latter are far more numerous in the human brain and their variability in different individuals tends to obscure still further the pattern of the primary sulci (Radinsky, 1974).

Size alone is not the only significant attribute when considering the evolution of the brain, since variation in the number of cell layers may be associated with the degree of neural interconnections and associations. Furthermore, there is no palaeontological evidence to indicate that the brain began to expand to modern dimensions before the latter part of the Early Pleistocene, although its subsequent expansion was rapid. Thus, for several million years, the earlier representatives of hominoid evolution continued to retain a brain of simian dimensions. Hence, while a large brain may be considered diagnostic of the genus *Homo*, it cannot be regarded as a criterion of the Hominidae as a whole. So far as meager evidence indicates, it seems likely that the pongid type of brain was by no means complete in its evolutionary development by early Miocene. Furthermore, the endocranial casts of australopithecines do not illustrate a marked divergence in their general shape from existing apes, nor do their surface markings provide any great differences—although contrary views have been expressed by Broom and Schepers (1946). It is evident, therefore, that although there have been marked changes during the evolution of the primate brain, purely qualitative (subjective) information must be supplanted by quantitative (objective) data, if any insight is to be obtained.

METRICAL DEFINITION OF BRAIN SIZE

Determination of the capacity of the endocranial cavity due to its complexity and foramina is difficult so that when specimens are incomplete, e.g. fossil specimens, the determination must, to a certain extent, be subjective. There is abundant material of extant species for the relationship between brain size and behavior to be studied directly, although estimation of the endocranial capacity from dried skulls is difficult since no data can be obtained concerning the actual brain tissue itself.

If the actual brain is not available, endocasts provide the only direct record of the evolution of the brain, and these can furnish information concerning the total volume, linear measures which can be combined to provide data relating to shape, including estimates of portions of the surface relative to the whole surface area: The diameter of nerve bundles can be deduced from the size of the cranial foramina. In many mammals, including lower primates, the inner surface of the neurocranium reflects the detailed configurations of the brain surface. In pongids and hominids, however, interpretation of the details of the highly convoluted brain surface from the endocranial cavity is difficult, since, in the living, the brain is surrounded by variable membranes, blood, and cerebrospinal fluid. Nevertheless, in fossil specimens, the only data relating to the brain is that available from the endocranial cavity, and even this may be incomplete or distorted.

Both natural and artificial endocasts provide impressions of the endocranial cavity. In the laboratory, the capacity of the endocranial cavity can be ascertained by filling the cavity with a relatively sightly compressible medium, such as millet seed (Breitinger, 1936) or water, although the more accurate impression materials used in dentistry have yet to be exploited in this respect. In all such determinations, however, cognizance must be taken of the fact that the brain is surrounded by variable amounts of tissue, so that estimations of the endocranial capacity can only be approximate.

In living specimens, the weight or volume of the brain itself can be measured, although here again intrinsic errors may arise from the manner of brain dissection, time of dissection following death, or method of preservation for the brain. Indeed these factors coupled with sampling and mensurational errors indicate that evolutionary changes must be interpreted with caution.

Endocasts of all genera of living prosimians reproduce all of the cerebral sulci (Bauchot and Stephan, 1967), and in the New World monkeys, endocasts preserve fine cerebral details in most of the genera, but cerebral details are blurred in the largest-brained forms, such as *Cebus* and the atelines. Only the smallest of the living Old World monkey endocasts clearly preserve the sulcal pattern but exceptional skulls of even the largest (baboons) show all sulcual details. Cerebral convolutions are reproduced on some gibbon casts, but most details are lost on all great ape and human endocasts (Connolly, 1950). The widely publicized lack of sulcual detail on human and great ape endocasts (Symington, 1916) may have been responsible for the general neglect of this area of study by primatologists.

Even where details of cerebral morphology are indistinct, endocasts reveal the relative size of some major parts of the brain, such as the cerebellum and occipital lobes, and allow general functional interpretation. For instance, relatively reduced olfactory bulbs and expanded occipital lobes in the endocast of a 55-million-year-old prosimian *Tetonius homunculus* indicate reduction of olfaction and increased importance of vision at least that far back in primate evolution (Radinsky, 1967). Similarly, comparison of endocasts of modern gibbons and monkeys reveals a relatively larger cerebellum in the former, which suggests more highly developed muscular control in gibbons compared with monkeys.

Recently, exploiting cranial associations, Olivier and Tissier (1975) have devised a series of formulae whereby the cranial capacity of a skull may be computed when only incomplete specimens are available. However, the correction factors to be applied to such computations depend upon knowledge of the taxonomic position. Nevertheless, when the cranial capacity and the measurements of an early fossil are known, their relationships make it possible to specify its taxonomic position and to distinguish *Homo erectus* from Neanderthal man.

Age Changes in Brain Size

As shown in Table I, there is a marked increase in brain volume with increasing age, particularly during the initial stages of development, and these are mirrored by changes in the endocranial capacity. The values listed in Table I refer only to modern man and there is no comparable data relating to fossil hominoids. In addition, there is as yet no data relating to the age changes in the nonneural contents of the endocranium with increasing age.

In man, brain growth appears to be complete at about 20 to 21 years (Pffister, 1903), although a few workers would extend

TABLE I
CHANGE IN SIZE OF BRAIN AND BRAIN CASE IN *HOMO SAPIENS* WITH INCREASING AGE*

Age	Volume of Brain	Volume of Brain Case	Percentage Volume of Brain / Volume of Brain Case
Birth	330	350	94.3
3 months	500	600	83.3
6 months	575	775	74.2
9 months	675	925	73.0
1 year	750	1000	75.0
2 years	900	1100	81.8
3 years	960	1225	78.4
4 years	1000	1300	76.9
6 years	1060	1350	78.5
9 years	1100	1400	78.6
12 years	1150	1450	79.3
20 years	1200	1500	80.0

* From *Tabulae Biologicae*, 1940.

the growth period somewhat (Zuckerman, 1928). In addition, certain disease processes may interfere with this growth pattern and certainly a reduction in brain tissue occurs with advanced age, although no such changes have been reported in the endocranial capacity. At the time of eruption of the first permanent molar, the endocranial capacity has achieved 94 percent of adult values in *Pan* (Schultz, 1940), 90 percent in *Gorilla* and 92 percent in *Pongo* (Ashton and Spence, 1958) and 94 percent in man (Todd, 1933). Weidenreich (1941) concludes that in both man and *Pongo*, growth of the endocranial cavity ceases at the time of third molar tooth eruption.

Despite these obvious age changes, however, there is frequently little comment about the age of the specimen when determining the endocranial capacity of fossil specimens, and it is difficult to predict future adult endocranial capacities from immature specimens (Sholl, 1948).

The Cerebral Rubicon

As shown in Table II, III, IV, and V there is considerable overlap between the endocranial capacities of hominoids and pongids, although some of this variation reflects the technique of measurement or the age of the specimens.

The limited data suggests that siamangs have a mean capacity 25 percent greater than gibbons with the means for males of most

TABLE II
VARIABILITY OF HOMINOID CRANIAL CAPACITIES*

Species and Series	Size of Sample	Sample Mean (cc)	S.D. of Mean	Estimated Coefficient of Variation	Reference
Hylobates agilis					
Combined males and females	21	98.8	10.32	10.45	Schultz, 1933
Hylobates lar					
Males	95	104.0	7.51	7.22	Schultz, 1944
Females	85	100.9	7.81	7.69	"
Symphalangus syndactylus					
Males	23	125.8	12.95	10.29	Tobias, 1971
Females	17	122.8	13.09	10.66	"
Pantroglodytes					
Males	24	420.0	33.33	7.94	Selenka, 1899
"	34	399.5	40.61	10.17	Zuckerman, 1928
"	33	410.0	47.70	11.63	Ashton and Spence, 1958
"	56	381.0	35.29	9.26	Schultz, 1965
Females	26	390.0	32.83	8.42	Selenka, 1899
"	27	365.8	33.70	9.21	Zuckerman, 1928
"	78	380.0	35.10	9.24	Ashton and Spence, 1958
"	57	350.0	28.85	8.24	Schultz, 1965
Pongo pygmaeus					
Males	36	434.1	50.86	11.72	Gaul, 1933
"	30	415.0	37.60	9.06	Ashton and Spence, 1958
"	57	416.0	36.44	8.76	Schultz, 1965
Females	59	389.8	36.27	9.31	Gaul, 1933
"	18	370.0	34.90	9.43	Ashton and Spence, 1958
"	52	338.0	32.89	9.73	Schultz, 1965
Gorilla gorilla					
Males	50	510.0	37.78	7.41	Selenka, 1899
"	22	505.4	43.19	8.55	Oppenheim, 1911–12
"	133	543.0	50.00	9.23	Randall, 1943–44
"	63	550.0	61.90	11.25	Ashton and Spence, 1958
"	72	535.0	71.13	13.30	Schultz, 1965
Females	48	450.0	33.56	7.46	Selenka, 1899
"	78	461.0	47.70	10.34	Randall, 1943–44
"	50	460.0	35.20	7.65	Ashton and Spence, 1958
"	43	443.0	39.41	8.90	Schultz, 1965
Homo Sapiens					
Males	1,000	s1345.0	169.00	12.57	Tobias, 1971

* After Tobias, 1971.

gibbon specimens centering around 100 to 104 cc, although *Hyplobates klossii* may have a somewhat lower capacity (Schultz, 1933). The overall mean for *Pan troglodytes* is 400 cc for males and 375 cc for females, where the small sample for *Pan paniscus* is 10 percent less for both sexes. For *Pongo*, the range of means

TABLE III
CRANIAL CAPACITY OF HOMINOIDS*

Specimen	Mean	Range
Australopithecus	500	405–600
Homo habilis	639	510–770
Pithecanthropus	880	700–1060
Sinanthropus	1075	845–1305
Modern man	1370	1070–1670
Classical Neanderthal	1470	1145–1795
A. africanus	Tobias, 1971	Holloway, 1970
1 from Taung (adult value)	540	440
Sts 60	435	428
Sts 5	480	485
Sts 71	480–520	428
Sts 19/58	530	436
MLD 37/38	480	435
Mean	494	442
A. robustus	Brain, 1968	Holloway, 1970
JK 1585	475	530

* After Tobias, 1971.

for males is 395 to 455 cc (Selenka, 1899; Oppenheim, 1911-1912) and 338 to 390 cc for females (Selenka, 1899; Schultz, 1965), so that the degree of sexual dimorphism is greater than in *Pan*. Finally, sample means for *Gorilla* are 497 to 511 cc for males and 442 to 478 cc for females, with the maximum value ranging from 605 cc (Gyldenstolpe, 1928) to 685 cc (Randall, 1943-1944).

Many workers have recognized a "rubicon" of endocranial capacity between man and apes, which Keith (1948) puts at 750 cc and Vallois (1954) at 800 cc. The notion of a cerebral "rubicon" has been rejected by a number of workers (e.g., Straus, 1953). This is illustrated by the fact that the endocranial capacity of the largest gorilla is 752 cc which differs only marginally from the smallest Indonesian *Homo erectus* at 775 cc (Schultz, 1962). Indeed, Clark (1964) is at pains to emphasize that there should be no undue reliance placed on endocranial capacity in taxonomy. This again is illustrated by the fact that there are marked differences in the range of maximum endocranial capacity in male Gorilla: 752 cc (Schultz, 1962); 605 cc (Gyldenstolpe, 1928); 623 cc (Weidenreich, 1943); 652 cc (Harris, 1926); 655 cc (Bolk, 1925) and 685 cc (Randall, 1943-1944). Moreover, Schultz (1957) and others have stressed that absolute brain size is of less phylogenetic importance than brain size relative to body

TABLE IV
CRANIAL CAPACITY OF *HOMO ERECTUS*

Site	Specimen		Volume*	Reference
Indonesia				
	H. erectus erectus	I	850	Tobias, 1967
		II	775	Weidenreich, 1943
		III	890	Weidenreich, 1943
		IV	750	von Koenigswald, 1967
		VI	975	Jacob, 1966
		VII	915	Sartono, 1968
China				
	H. erectus pekinensis	II	1030	Weidenreich, 1943
		III	915	"
		X	1225	"
		XI	1015	"
		XXI	1030	"
	H. erectus subsp.		780	Woo, 1965
Tanzania				
	H. erectus subsp. (Olduvai hominid 9)		1000	Tobias, 1965

* ∴ sample range 750–1225 cc mean 935 cc

TABLE V
CRANIAL CAPACITY OF *HOMO ERECTUS*

Specimen	Range
Homo erectus erectus mean of 5 = 848 cc*	750–975
Homo erectus of Choukoutien mean of 5 = 1043 cc	915–1225
Homo erectus of Lantian 1 = 780 cc	
Homo erectus of Olduvai 1 = 1000 cc	

* Mean of 12 = 936 cc

size: The variability of endocranial capacities of gorilla may merely reflect variation in body weight.

If logarithms of means against standard deviations of endocranial capacities of apes and modern man are plotted, a linear relationship emerges. If fossil hominids are subsequently inserted, then an exponential relationship emerges, which shows an explosive acceleration of hominization, with no regularity in the order of appearance of hominoid species (groups). Thoma (1969) has demonstrated that a reduction in endocranial capacity occurred in recent times. For instance, this worker's estimate of males in the Upper Paleolithic at about 1600 cc, and since the endocranial capacity of present-day man is 1350 cc, a 10 percent reduction in the cranial capacity is evident. It is possible that this reduction in endocranial capacity is the result of cultural and instrumental adaptation.

Great care must be exercised however in the interpretation of evolutionary changes, not only in relation to the endocranial capacity, but also with other morphological features. On the one hand, the hominid evolutionary lineage still requires further elucidation (Pilbeam, 1969). On the other hand, there is conflict in the endocranial capacity determined by different workers for fossil specimens (Holloway, 1970; Tobias, 1971).

Examination of Table VI shows that in general, large mammals have large brains, e.g. *Pan* has a larger brain than *Macaca*, although there are obvious exceptions to this general rule. For instance, the cranial capacity of *Homo sapiens* is twice that of *Gorilla*. The human cranial capacity, however, varies greatly, since as shown by Clark (1959), the normal limits are 900 to 2,300 cc. Nevertheless, although female *Homo sapiens* may have a brain weight of 1,200 gm, this is large compared with that of a rhinoceros (60 gm). Hence, brain size may be more related to

TABLE VI
RATIO OF BRAIN WEIGHT: BODY WEIGHT*

Specimen		Ratio
Squirrel monkey	(*Pithisciurius sciurius*)	1 : 12
Marmoset	(*Leontocebus geoffrey*)	1 : 19
Japanese mouse		1 : 22
Porpoise	(*Phocaena communis*)	1 : 38
House mouse	(*Mus musculus*)	1 : 40
Tree shrew	(*Tupaia javonica*)	1 : 40
Man	(*Homo sapiens*)	1 : 45
Ground shrew	(*Sorex minutus*)	1 : 50
Monkey	(*Macaca rhesus*)	1 : 170
Gorilla	(*Gorilla gorilla*)	1 : 200
Elephant	(*Eliphas indicus*)	1 : 600
Whale	(*Phseter catadon*)	1 : 10,000

* After Cobb, 1965.

the complexity of the tasks it has to perform rather than to the size of the body it has to control. Lilly (1961) believes that a brain weight over 1,000 gm is necessary for language and "intellect," although normal *Homo sapiens,* in particular small Asiatic women, show perfect language ability and good intellect and yet possess brains weighing less than 1,000 gm (Shafer and Symington, 1909). It would appear, therefore, that there is no "critical" myth of 1,000 gm brain weight for intellect (Hechst, 1932), especially since human nanocephalic dwarfs with brains less than 700 gm can talk and converse fluently (Lenneberg, 1964).

Rather than brain weight in isolation, other workers see significance in the relationship between brain weight and body weight (Table VI). This relationship shows that man and porpoises are linked together, but separated by a mouse and a shrew, which contrasts with a squirrel monkey and Japanese mouse, who have relatively larger brains than *Homo sapiens.* Thus, little constructive information is evident from the ratio of brain weight to body weight. Also, as there is considerable variation in brain weight within a species, it is difficult to compute means with any degree of accuracy. For instance, the average human brain weighs 1000 to 2000 gm (Cobb, 1965), 850 to 2100 gm (Clark, 1959), or 700 to 1700 gm (Donaldson, 18895). The brain also suffers from atrophy with increasing age and many other physiological and pathological factors, and there is little definitive data concerning the brain weight variation in nonhuman primates. In addition, brains differ in their compactness, such that the interpretation of brain weight alone may obscure relevant facets of neural function.

Another approach in the evolution of brain size has been devised by Jerison (1963) who derived a coefficient to estimate the "efficiency" of brains which is related to the total volume of neurones in the cortex on the basis of brain weight. It is clear, however, that the question has yet to be resolved as to whether the brain should be weighed with or without cerebrospinal fluid, fresh or fixed, both being features which may provide considerable variation in brain weight.

If brain weight is related to body size, then the human brain is just over 2 percent body weight, while that of the house mouse is 2.5 percent (Table VII). This places man at a disadvantage

TABLE VII
BRAIN WEIGHT*

Specimen	Weight in gms
Whale	7000
Elephant	5000
Porpoise	1700
Man	1400
Woman	1300
Rhinoceros	600
Gorilla	500
Macaque monkey	90
Cat	30
Mouse	0.4
Weight of certain "normal" human brains	
Australian bushwoman	794
Anatole France	1017
Japanese woman (average)	1250
Walt Whitman	1282
European woman (average)	1300
European man (average)	1400
Thackeray	1658
Bismark	1807
Cuvier	1830
Daniel Webster	1895

* After Cobb, 1965.

relative to other animals, so that as a consequence, other relationships have been devised to assert his supremacy! For instance, when hypothalamic length is expressed as a fraction of that of the cerebrum man has the lowest value relative to other animals, and so is distinguished (Kummer, 1961). Other indices have also been determined, e.g. the weight of the spinal cord is a fraction of brain weight (Latimer, 1950) or the endocranial capacity relative to the area of the foramen magnum (Radinsky, 1967), where in both instances, man reigns superior from other mammals.

There are other variables relative to brain weight. For example, Pearl (1905) reports that brain weight in man is a function of body weight and stature, although others consider that brain weight is affected by body height not body weight (Pakkenberg and Voigt, 1964; Spann and Dustmann, 1965). This only illustrates that brain weight is not the only important factor in elucidating human evolution, especially when it is apparent that the method of brain dissection from a skull will have a direct bearing on the final weight of the brain. Furthermore, although the volume of the brain can accurately be determined from living species, no clue of the brain weight can be ascertained from fossil specimens.

Variation in Cranial Capacity

Rensch (1960) considers that man has great selective advantage over other mammals in the increased brain size; special complexity of the nervous system associated with division of labor, particularly with the development of the speech area of Broca; the prolongation of the juvenile period allowing time for learning while still subject to parental protection; erect posture, freeing the hands for tool use and skills. While agreeing with the majority of these points, Cobb (1965) believes that the most important neocortical area in evolution is the posterior parietal region of the cerebral cortex with its probable functional capacity to synthesize sensory output into concepts and symbols. This worker considered that Broca's area, although essential, is part of the motor area associated with speech and so furnishes a less crucial role in the evolution of thought. Obviously, more information is required before the most critical area of the neocortex in hominoid evolution can be established.

In hominids at least, the endocranial capacity is but an outward manifestation of multiple internal organizational changes that are better suited to correlation with behavioral variables.

Comparisons based on endocranial capacities alone are not comparisons of equal units, and indeed, the extreme variability of this parameter without demonstrated correlation or causal relationship to behavioral parameters undermines its effectiveness in discussing human evolution. This position does not mean, however, that this parameter is useless, but much more reliable data on endocranial capacities are required. It is a good parameter for characterizing morphological variability and for providing a statistical basis for taxon comparison and elucidation. If enough data were available, then the endocranial capacity might be sensitive enough to distinguish between variability attributable to sexual dimorphism, subspecific or specific variation. It could provide a basis for reasonable speculation concerning selection dynamics. Endocranial capacities, particularly in the fossil record, might even be used to estimate body weights and heights, parameters useful for analyzing past ecological relationships such as carrying capacity.

An increase in brain size can be interpreted in a number of ways, including more nerve cells, bigger nerve cells, more neuroglia, more nerve processes, thicker nerve processes (myelin sheaths) or more highly branched nerve processes. Obviously, gross brain size alone cannot explain differences in behavior within the primate order (Holloway, 1968). Nevertheless, the bigger the brain, the lower the density of nerve cells (Tower, 1954) and so the more space for synapses and so more complex behavior. There is however, no method of establishing the neuronal density of fossil specimens.

Witherspoon (1960) computed a correlation between cranial capacity and motor development and suggested that it was possible to predict behavior patterns of fossil primates. This worker correlated brain growth curves (endocranial capacity) for monkey, chimpanzee, and humans to ten behavioral categories, based on the assumption that motor and brain development are related. Nevertheless, whether or not there is an accurate correlation between endocranial capacity and brain development has yet to be elucidated critically, since the ontogenesis of motor development is related to maturation of subcortical nuclei and progress in myelinization, which are not necessarily associated with endocranial expansion. Using this assumption, however, Witherspoon found that the ten behavior categories appeared in much the same order in chimpanzee and man.

Witherspoon (1960) noted that it should be possible to construct a generalized curve for primates reflecting age, behavioral and percent adult brain weight variables, although he did not actually construct such a curve. It must be emphasized, however, that a large number of endocasts would be required to compensate for variations within fossil species. Also, due to the lack of data, it would not be possible to ascertain the behavioral attributes of fossil specimens since subcortical components and cortical-subcortical interactions determine behavioral patterns: features possibly unrelated to endocranial capacity.

Jerison (1963) considered that brain size was directly related to size of the body and indirectly related to body size, i.e. surplus neurones, which are present over and above those required for the satisfaction of immediate bodily needs. These "surplus cells" effect the intelligent adjustment to the environment. On the basis of cell counts in a variety of primates and given certain assumptions, Jerison claims that it is possible to estimate the number of cortical nerve cells not only in the brain as a whole, but also in each of the two components. He has developed a series of equations for the calculation of these neuronal values, given brain and body size. By applying these formulae, he has been able to compute the number of "extra" neurones regarded as being available for brain-behavior adaptive mechanisms. With this second parameter (the number of excess neurones) he has found it possible to differentiate various primates and especially *Homo sapiens,* on the basis of the number of extra neurones, e.g. modern man has far in excess of *Pan* and *Gorilla* (see Table VIII).

The most interesting feature to emerge from this table is the way in which *Australopithecus africanus* and *Zinjanthropus* are distinguished from the anthropoid apes, despite the fact that their absolute brain size is in the anthropoid range. It is clear that the australopithecine level is to be considered divergent from the anthropoid line in terms of brain development, although the divergence is relatively small.

TABLE VIII
ESTIMATES OF "EXTRA NEURONES" IN HOMINOIDS*

	Endocranial Capacity (cc's)	Estimated Body Size (cc's or gms)	Estimates of total neurones (in thousands of millions)	Estimates of body related neurones (in thousands of millions)	Estimates of "extra neurones" (in thousands of millions)
Chimpanzee	460	45,000	4.3	0.9	3.4
Gorilla A	540	200,000	5.3	1.8	3.5
Gorilla B	600	250,000	5.7	2.1	3.6
A. robustus	500	45,000	5.0	0.9	4.1
A. africanus	500	25,000	5.0	0.7	4.3
H. erectus (range)	750–1,225	50,000	6.6 – 9.4	0.9 – 1.0	5.7 – 8.4
H. sapiens	1,300	60,000	9.5	1.0	8.5

*After Sevison, 1968; Tobias, 1971.

There are certain criticisms regarding such an analysis:

1. The histological foundations of these assumptions require strengthening;
2. there is almost complete absence of data relating to the differences in the numbers of cortical neurones among individuals of the same species and the apparently small differences between closely related species that have similar brain size;
3. there is no information whether the neurone number—brain size relationship computed for contemporary animals—holds true also for fossil forms; and, finally,
4. it does not adequately take into account regional variations in neurone number, the ratio of neurones, neuroglia, size of nerve cell bodies, and the length and complexity of dendritic processes.

Thus, any conclusions must be regarded as tentative for the present. Holloway (1968) criticizes Jerison's method on the basis that the mathematical extrapolations are based on assumptions of cortical volume and neuronal density and overlooks the fact that in primate evolution the cortex has undergone reorganization. Indeed, several workers have recently stressed the major reorganization of the internal structures of the brain which must have occurred in evolution (Holloway, 1969; Bonin, 1963).

In general, it is assumed that the evolution of the primate brain has involved only the addition of more units (Garn, 1963), although obviously there were also minor changes in the somatosensory limbic and other brain subsystems. Thus, the human brain is not simply an enlarged version of an ape or monkey brain and many changes have possibly occurred. For instance, there may have been an increase of "association" cortex at the expense of sensorimotor cortex during primate evolution (Bonin and Bailey, 1961) and increase occurring mainly in the parietal and temporal lobes (Weidenreich, 1936). The use of the average cortical density by Jerison ignores this process of reorganization and, of course, endocasts provide little data on this matter. There is also the problem of corticalization of function, which relates to the fact that damage to the motor or sensory cortices has increasingly detrimental and more permanent effects as the phylogenetic scale is advanced (Noback and Moscowitz, 1962). Again, such corticalization is difficult to quantify in terms of neuronal number. Furthermore, recent physiological research has made it increasingly clear that the neural unit does not simply comprise a neurone, but the neurone coupled with its neuroglial companions (Hyden, 1962). Nevertheless, since there is an increase in general neurone size with increasing brain size (Bonin, 1938) and the glia are necessary components to the metabolic functioning of the neurone, it seems most probable that glial cells are related to brain size rather than phylogenetic level, i.e. a large component of cortical expansion may be glial cells. Only more comparative studies are necessary to elucidate, therefore, how far cortical expansion is related to neuronal or glial numbers.

Holloway (1966) has shown that the human microcephalic (capacity less than 600 cc) is sometimes capable of the human specificity, i.e. these people appear to interact with their environments in the human manner with less cortical neurones than a healthy chimpanzee. This underlines the fact that the neural structures involved in the human specificity have some locus in terms of their organization or interactions which overrides any single mass parameter, e.g. cranial capacity or neuronal number. Thus, the question of the mental evolution of hominoids must always rest on indirect evidence, and so must await the acquisition of more reliable data.

ENDOCRANIAL CAPACITY AND CULTURE

Cultural advancement is as important a nonmorphological feature of hominization as increase in endocranial capacity is a morphological characteristic. The capacity for tool making is widespread among great apes and in lower primates (Kortlandt and Kooij, 1963; Hall, 1963; van Lawick Goodall, 1968; Beck, 1972), and these workers report that, unlike monkeys, great ape behavior is largely molded by maternal education, social traditions, and other environmental factors rather than innate factors. According to these workers, therefore, great apes and man may be designated as "cultural Primates," while the gibbons and man may be classified as "instinctural Primates." There is, therefore, a tendency among primatologists to assign great apes with man because the ape behavioral pattern is more learned than genetic or instinctural. Rather than there being two discrete categories of behavior, however, in fact there is a smooth transition from the instinctural to the learned patterns of behavior. Even concerning modern man, there is doubt as to the degree to which instructural patterns have been discarded for learned ones.

Tool behavior was undoubtedly a critically important factor in hominid evolution and radiation (Washburn, 1960), although Lancaster (1968) argues that tool use was present even before the hominid grade was attained. Beck (1972) supports this conclusion, since tool use is possible with sensory, motor, and cognitive capacities no more specialized than that of extant baboons.

Although the mean endocranial capacity of *Australopithecus* is similar to that of *Gorilla*, there are sufficient postcranial remains to indicate body weight less than that of *Gorilla*, i.e. the estimated brain/body weight ratio (relative brain size) of *Australopithecus* exceeds that of the biggest-brained species of the living great apes. Furthermore, the external configuration of the australopithecine brain shares many man-like features. Moreover, it would seem likely that a brain that is more hominized in size and shape is also more hominized in its fine internal structure necessary to facilitate the emergence of more complex behavioral patterns than those of the apes. There is also direct evidence to confirm the complex tool-making capacity of these early hominids (Dart, 1957; Brain, 1970).

A preponderance of cateodontokeratic activities need not have characterized all early hominid populations of the Lower Pleistocene. The South African australopithecines of the Lower Pleistocene may have been atypical. The bone artifacts of Makapansgat may even represent the persistence in South Africa of a yet earlier, possibly Mio-Pliocene phase of cultural hominization. Indeed, Pilbeam and Simons (1965) have reported direct and indirect evidence of small canined, bipedal, and possibly tool-using hominids in the late Miocene or early Pliocene. It is true that these hominids were not all ancestral to the hominids known from the Pleistocene, but represented a consequence of a quite independently initiated hominizing trend within one or more species of Tertiary catarrhines. Such a notion is compatible with the possibility that hominizing tendencies could have recurred several times in the history of other Primates as indicated by the evidence of Leakey (1974).

There is evidence to suggest the following stages in cultural evolution: pongids—sporadic, unsystematic implemental activity of dubious survival value;

Australopithecus: more versatile implemental activities, incipiently systematic and with a definite survival value;

Homo erectus: sustained cultural development, comprising consistent patterns, diversified tool types and the use of fire, high survival value of culture;

Homo sapiens: all significant trends of cultural hominization intensified, so that man is absolutely culture-dependent for survival.

Some characteristics, e.g. hunting, are difficult to place in this taxonomic system, although by the time of *Homo erectus* emergence, hunting had become a systematic obligatory mode of life, and mutual attributes necessary for sustained predation must have appeared. Krantz (1968) has stressed the relationship between brain development and hunting revolving around the development of memory. If this notion is correct, then an increase in endocranial capacity of 500 cc in *Homo erectus* compared with *Australopithecus* must have resulted in increase in memory. Moreover, a brain at least two-thirds the size of that of modern man's should have permitted *Homo erectus* to engage in hunting in a manner approaching that observed in recent man. Nevertheless, there is no conclusive proof of an association between brain size and memory.

Krantz (1961) suggested another mechanism of selection for increased brain size from *Australopithecus* to *Homo erectus*, based on the growth pattern of the hominid brain. This worker suggested that the small-brained Palaeolithic people were as well endowed as adults as their modern descendants; but that as young children they were not capable of using symbolic language.

This would, therefore, reduce the time available for acculturation and so limit the culture content. Krantz accepts that "virtual humanity" is reached with a normally developed brain in excess of the 750 cc threshold. For the modern child, 750 cc is the approximate endocranial capacity at the end of the first year of life (a time coinciding with the initiation of symbolic language), so that 750 cc is a threshold. Although this does not guarantee speed if the nervous system is not adequately developed, it represents a threshold which provides a lower limit below which symbolizing is not possible. Krantz subsequently produced a hypothetical curve of brain growth for *Homo erectus*, setting the curve at a constant 61 percent of the *Homo sapiens* curve. The curve showed that the endocranial capacity of *Homo erectus* did not exceed the threshold of 750 cc until after the sixth year of life, which would limit the mental abilities of these children. This thus reduced the period of full cultural participation, thereby limiting the total quantity and complexity of cultural content that is likely to be transmitted to succeeding generations. Logically, therefore, when the age of onset of symbolizing is lowered by increasing the child's brain size, the quantity of cultural material that can be transmitted is increased.

This would account for the slow rate of change in the lower Palaeolithic and may explain the large mass of superfluous brain tissue in modern man; provided it is assumed that the pattern of brain growth would not have changed during evolution from *Homo erectus* to *Homo sapiens*. The pattern does change appreciably, however, between modern pongids and modern hominids (Zuckerman, 1928), the most notable distinction being the steep initial postnatal rise in man compared with apes. There may possibly have been a steady change in the growth curve from the pongid pattern to the modern human curve during the several stages of hominization. Nevertheless, the major changes are likely to have occurred from pre-Australopithecus to *Australopithecus* to *Homo erectus*, and this pattern of growth would possibly have remained essentially similar to that of modern man by the time of *Homo erectus*. This hypothesis has yet to be substantiated.

Washburn (1967) stresses that behavior precedes structure

in evolution, which denigrades the role of the brain during human evolution. The contrary view has been formulated by Holloway (1969), who argues that the brain enjoyed a primary role in human evolution. In fact, a blending of physical and cultural evolution probably occurred, although until this controversy is cleared up, the value of research into the endocranial capacity of hominids is difficult to substantiate.

BRAIN SIZE AND RACE

It has often been claimed that races differ in intellectual capacity, particularly when relating to the intellectual inferiority of Negroid populations (Swan, 1964; Putnam, 1967). For instance, Swan (1964) states that the average cranial capacity of European whites is from 100 to 175 cc greater than that of African Negroes, and their average brain weight is some 8 to 12 percent greater. Only a proportion of the endocranial capacity comprizes brain tissue and estimates of this proportion vary from 10 percent (Brandes, 1927) to 33.33 percent (Mettler, 1955). Furthermore, the ratio varies in an adult lifetime depending upon age and certain illnesses, although the endocranial volume itself does not alter.

In 1909, Mall reported that the average Negro brain weight is about 100 gm less than that of whites, and Bean (1917) also noted lower brain weight in Negroes than whites. Pearl (1934) stated that the mean weight of 139 brains of Negroes was 1354.8 gm, which is 92.1 percent of the mean of twenty-four brains of whites (1470.6 gm.). These facts have been subsequently used by Putman (1963) to indicate a lower intelligence of Negroes. Nevertheless, these values are averages, whereas the ranges of brain weight actually overlap one another considerably.

Matiegka (1902) and Pearl (1905) claimed that in man brain weight varies with body weight and body height, i.e. taller and heavier people have larger brains. More recently, Pakkenberg and Voigt (1964) have reported that brain weight depends significantly on stature but not on body weight, a feature confirmed by Spann and Dustmann (1965). Thus, clearly, if varying average body sizes in different population samples are not taken into account, little significance can be assigned to comparisons between brain weights. In addition to variation in stature, brain weight is also affected by advancing age (Appel and Appel, 1942) and nurtritional state (Brown, 1966). Also, brain weight in a population depends upon the sample actually measured, yet as indicated by Pakkenberg and Voigt, (1964) the majority of the data are based upon subjects derived from hospitals, including mental institutions. Brain size may also be affected by the time that has elapsed between time of death and removal of the brain and the technique of actual dissection of the brain from the neurocranium. Finally, cognizance is rarely taken of the fact that variation in brain weight may occur within the limits of normal functioning, and many workers have reported little correlation between brain weight and mental capacity (Bonin, 1950; Holloway, 1968). Moreover, if brain weight is correlated with mental ability, then the future is bleak. This is because there has been a reduction in endocranial capacity from the Neanderthal period to the present day, and there is no sign of this diminution decreasing. This may be accounted for by the fact that selective pressures, which once placed a considerable premium on big brains, have been somewhat relaxed, so that brain size is no longer the yardstick to survival as it may once have been.

While it is commonly considered that there is no relationship between brain size and intelligence in modern man, a casual relationship is usually granted when different species are concerned. Nevertheless, Van Valen (1974), from reanalysis of craniometric data obtained prior to the 1930s, suggests that there is a relationship in man strong enough to be of major evolutionary importance. Moreover, Bielicki and Welon (1964) and Huizinga and Slob (1965) found a relationship between sib number (an approximate index of fitness) and cephalic index, whereas Clark and Spuhler (1959) report a relationship between head length and fertility and Falconer (1966) have described a relationship between fitness and intelligence quotient. Most of the material inside the human cranium is not, however, directly associated with thought (Tobias, 1970), although the amount of this material must, to a certain extent, be related to intellectual function and its preservation. Also, the work of Lashley (1929) and Pietsch (1972) shows that most of the brain of at least some vertebrates is redundant, i.e. an increase in size or improvement of construction of a brain might not cause more effective thought. Nevertheless, Lashley (1949) noted that the "severity of deterioration (of behavior) is proportional to the quantity of cerebral tissue destroyed" while Braestrup (1971) reports a genetic correlation between brain size and intelligence, i.e. some genes or equivalent genes affect both characters. Brain size has increased dramatically and at an unusually high rate during human evolution (at least until Neanderthal times), and it is possible that this may be related to intelligence. Furthermore, it is likely that in the hunter-gather populations whose survival was associated with the availability of food and shelter, intelligence may be relevant to fitness for survival, and so have evolutionary significance, although further data are required in order to support this supposition.

Recent work in palaeontology suggests that there is an increasing use of a combination of archeology, skeletal, and primate evidence in interpreting the possible ranges of behavior during human evolution. Physical anthropologists are initiating serious investigation of neurological evolution in much the same way that the field has approached the evolution and adaptability of the musculoskeletal system, blood group genetics, and human adaptability. Thus, the physical anthropologist may be in the interesting position of integrating his findings with those of the social, cultural, and linguistic anthropologists in investigating some important biobehavioral problems that are fundamental to all anthropology. Nevertheless, as evident throughout this chapter, there is a dearth of data relating to the endocranial capacity of Primates. Hence, the interpolation of evolutionary significance must await the acquisition of more data.

The size of the brain, considered in conjunction with differences in body weight by aid of the allometry formulae, is not more pronounced in Primates than in other orders. A prosimian stage reached by many mammals and a Simian or even Pongid stage is matched by several semiaquatic and/or aquatic groups. Only man has encephalization which exceeds that of all animals and represents the only Primate with an outstanding brain size. Brains of fossil forerunners of man represent intermediate stages and fill the gap between man and recent nonhuman Primates.

The extraordinary brain size of *Homo sapiens* seems to have developed gradually.

The enlargement of the brain is not proportional in that all parts do not develop at the same rate. The neocortex is by far the most progressive structure and therefore used to evaluate evolutionary progress. Assuming that neocortical development has never been retrograde during phylogeny, the degree of neocorticalization is considered to be an effective guide in the discussion of phylogenetic sequences.

Directed size alteration of the nonneocortical structures in the ascending primate scale allows conclusions on prevailing trends in their phylogenetic development. Most of the structures or structural complexes so far investigated show progressive trends. The olfactory structures, in contrast, exhibit regressive trends. In several small structures, the trend can revert, so that a progressive phase in Prosimians may be followed by a regressive one in Simians.

On the basis of such prevailing trends, much data relating to the composition of a fossil brain and the progression of the various parts can be estimated when brain and body size are known. Yet such data can only be inferred, rather than established as fact.

Moreover, while the development of various cortical areas is a major determinant of external brain morphology, the relative size and shape of other parts of the skull may also affect the sulcul pattern and brain shape. For instance, the lack of cortical impression in *Daubentonia* and the long-skulled giant lemuroids *Megaladapis* and *Palaeopropithecus* suggests a lack of rostrocaudal compression which may be correlated with the loss of opercularized sylvian sulcus in these genera (Radinsky, 1970). Another example is the more convex dorsal profile of the frontal lobe of *Loris* compared to the condition in *Nycticebus,* which appears to have resulted from the small size and the relatively large orbits of Loris.

CONCLUSION

When investigating primate evolution, there is controversy as to which parameters should be examined. This is particularly relevant when considering the evolutionary changes of the brain and neurocranium, where the ranges of variability of brain mass may bear no particular relationship to behavior within a species, e.g. man. Moreover, mass parameters completely ignore reorganization among the nuclei and tracts which mediate specific behavioral variables. Similarly neurone number as an index of behavior overlooks reorganizational changes in the brain system. It is doubtful whether behavioral differences between taxa can be reduced to neurone number until neurone number is further broken down in terms of their functional affinities. It is doubtful, however, whether even such breaking down would accommodate the role of subcortical structures in cortically mediate behavior. Hence, volumetric studies of the brain and investigations into neurone number are unsuitable for making anything but the most crude behavioral correlations—the same ones which we have faced for the past hundred years. A more specific parameter would be the proportion of the cerebral cortex involved in particular behavioral activities, but as yet there is no quantitative data on this topic. At a more microscopic level, there are paramatters such as cell size, neural density, glial/neural ratios, degree of dendritic branching, although there is still the question of how to relate these parameters to primate behavioral patterns.

The evolution of the brain cannot be viewed in isolation. For instance, the brain influences both the internal configuration and external form of the skull. In primates, the shape of the occipital region of the skull is influenced to an increasing extent by the cerebellar hemispheres and the occipital lobes of the cerebrum. Nevertheless, the reason why the endocranium copies negatively the surface of the brain in one case and not in another are unknown. It is obvious that the shape of the brain, especially the cortical gyri, can only be brought into connection with the configuration of the tabulae interna of the bones of the neurocranium if the space between the skull and the surface of the brain is very small. In those cases, the endocranial casts have a similar morphology to the external surface of the brain itself. In other animals, where the same topographical presuppositions pertain between the brain and the endocranium, no impressiones gyrorum are developed at all, e.g. in *Homo, Gorilla,* and *Pongo*. In *Pan,* the impressiones gyrorum are indistinct. It is possibly a physiological process in the later ontogenesis that results in the development of the impressiones gyrorum, although the process is unknown.

Identification of the selection pressures which resulted in a doubling of primate brain size between the basic condition of the prosimians and the appearance of the anthropoids remains controversial. The prosimian brain had rapidly reached the size of the "average" of living mammals early in the primate radiation, but the typical anthropoid brain is between two to three times the size of the prosimian brain. Furthermore, the limited fossil evidence suggests that the increase in anthropoid brain size occurred relatively early in their radiation, perhaps in the Oligocene. The adaptive advantages of the enlarged brains of the anthropoids may have been associated with superior forms of vision, including more complete stereopsis. Such adaptations with elaboration of the cortical representation of vision, hearing, and motor behavior in the anthropoids would have been associated with superior locomotion and perception, including the development of brachiation as a form of movement. This would have resulted in a much more arboreal mammal than had previously evolved.

Within the *Anthropoidea,* there is too little data about the evolution of the brain during the transition from monkeys to apes to make valid evolutionary judgements. An analysis according to the "extra neurones" hypothesis clearly placed the living apes above the monkeys with respect to the number of neurones related to "intelligent" behavior. Most students regard that the pongids are more capable of elaborate and more plastic behavioral adaptations than are either New or Old World monkeys. Nevertheless, present evidence is insufficient to permit conclusion about the transition between monkeys and apes to be applied to evolutionary analyses. It is easy to imagine the percursors of the hominids as derived from a simian as from a pongid stock if we are restricted to evidence of the present behavioral capacities of these groups.

The evolutionary progression in primates that culminated with the achievement of the brain size of *Homo sapiens* covered over ten million years according to present evidence. Moreover, it is possible that the relatively small-brained australopithecines were probably the typical hominids for most of the last five million years. So long, therefore, that we must assume there to have been an hominid adaptive zone without much enlargement of the brain. This has important implications for the type of model we construct of the selection pressures that resulted in the enlarged brain (*Homo erectus* and *Homo sapiens*) when this finally occurred. Also, it is possible that the enlargement of the human brain could have followed rather than preceded the invasion of that hominid niche. The fact that the australopithecines had survived as a genus for approximately four million years demonstrates that the selection pressures of the australopithecine niche were not of themselves sufficient to produce the enlarged brain that one associates with true men, although they probably forced some enlargement of the brain. The probable adaptations of the australopithecines must have served as precursors for later hominid adaptations by permitting hominids to explore the niches that were relatively advantageous for a species with a significantly enlarged brain.

The selection pressures that led to our enlarged brains must have appeared at the stage of human evolution represented by *Homo erectus*. It is in the contrasts between the australopithecines and the pithecanthropines rather than between the australopithecines and living forest apes or savanna baboons that we may identify the selection pressures that led to man, although there is no definitive evidence as yet. Nevertheless, since the evolution of this enlarged brain, only minor developments of the brain and intelligence can be attributed to organic evolution. The role of tool making is trivial in the analysis of the evolution of the brain, since it could be accomplished with very little brain tissue, and of itself would imply little of an enlarged brain. It is only if we assume that imagery was involved in tool making that we have reason to expect a selective advantage from a large-brained species. Comparable arguments would also hold for other categories, e.g. learning.

As evident from the review of the comparative anatomy of the Vertebrate nervous system (Kappers, Huber, and Cosby, 1936), further data are required before the evolutionary changes of the primate brain can be both identified and assessed quantitatively. This is also complicated by the fact that brain weight is correlated with body height (Spann, 1956), so that the interrelationship between these two variables may have some bearing on the evolution of the primate brain. Finally, from examination of the endocranial cast of the *La Chapelle-aux Saints* skull, Le May (1975) speculated that the neanderthaloid brain resembles that of modern man in an area important for speech. This suggests that the neanderthaloids possessed the neural development necessary for language, a feature which again requires further enquiry. Nevertheless, it may not be to modern civilizations that we must look for the main importance of a relationship between intelligence and brain size. Brain size may have stopped increasing by neanderthal times, despite the fact that brain size has increased dramatically, and at an unusually high rate during human evolution. It is likely that in human gatherer populations, whose density is ultimately or proximally regulated by the availability of food or shelter, intelligence has had some importance to fitness and evolution, but until some method of estimating intelligence of fossil specimens has been established, future work will remain speculative.

REFERENCES

Appel, F. W. and Appel, E. M.: Intracranial variation in the weight of the human brain. *Hum Biol, 14*:48, 235, 1942.

Ashton, E. H. and Spence, T. F.: Age changes in the cranial capacity and foramen magnum. *Proc Zool Soc Lond, 130*:169, 1958.

Bauchot, R. and Stephan, H.: Encephales et moulages endocraniens de quelques insectivores et primates actuels. Problemes actuels de paleontologie. *Colloq Inter Centre Nat Rech Sci, 163*:575, 1967.

Bean, R. B.: The weights of the organs in relation to type, race, sex, stature and age. *Anat Rec, 11*:326, 1917.

Beck, B. B.: Tool use in captive Hamadryas baboons. *Primates, 13*:276, 1972.

Bielicki, T. and Welon, Z.: The operation of natural selection of human head form in an East European population. *Homo, 15*:22, 1964.

Bolk, L.: On the existence of a dolichocephalic race of gorilla. *Proc K Ned Akad Wet, 28*:204, 1925.

Bonin, von G.: A study of cell size in cerebral cortex. 11 Motor area of man, oebus, cat. *J Comp Neurol, 69*:381, 1938.

Bonin, von G.: *Essay on the Cerebral Cortex.* Springfield, Thomas, 1950.
Bonin, von G.: *The Evolution of the Human Brain.* Chicago, U of Chicago P, 1963.
Bonin, von G. and Bailey, P.: Patterns of the cerebral isocortex. *Primatologia*, Bnd 11 Teil 2, Lief 10, 1961.
Braestrup, F. W.: The evolutionary significance of learning. *Vidensk Meddel Dansk Naturhist Foren*, 134:89, 1971.
Brandes, K.: Liquorverhaltnisse an der Leiche und Hirnschwelling. *Frankfurt Ztschr f Path*, 35:274, 1927.
Breitinger, E.: Zur Messung der Schadelkapazitat mit Senfkornen. *Anthropol Anz*, 13:140, 1936.
Broom, R. and Schepers, G. W. H.: The South African fossil ape-men, the Australopithecinae. *Transv Mus Mem*, No. 2, 1946.
Brown, R. E.: Organ weight in malnutrition with special reference to brain weight. *Devel Med Child Neurol*, 8:512, 1966.
Campbell, C. B. G.: The visual system of insectivores and primates. *Ann N Y Acad Sci*, 167:338, 1969.
Clark, P. J. and Spuhler, J. N.: Differential fertility in relation to body dimensions. *Hum Biol*, 31:126, 1959.
Clark, W. E. Le Gros.: The crucial evidence for human evolution. *Am Phil Soc Proc*, 130:159, 1959.
Clark, W. E. Le Gros.: *The Fossil Evidence for Human Evolution.* Chicago, U of Chicago P, 1964.
Cobb, S.: Brain size. *Archs Neurol Chicago*, 12:555, 1965.
Connolly, C. J.: *External Morphology of the Primate Brain.* Springfield, Thomas, 1950.
Donaldson, H. H.: *Growth of the Brain: Study of Nervous System in Relation to Education.* New York, Scriber's, 1895.
Falconer, D. S.: Genetic consequences of selection pressure. In Meade, J. E. and Parkes, A. S. (Eds.). *Genetic and Environmental Factors in Human Ability*, Edinburgh, Oliver and Boyd, 1966, pp 219-232.
Garn, S. M.: Culture and the direction of human evolution. *Hum Biol*, 35:221, 1963.
Geschwind, N.: Development of the brain and evolution of language. *Georgetown University Monogr Ser Lang and Ling*, 17:155, 1964.
Gyldenstolpe, N.: Zoological results of the Swedish expedition to Central Africa. *Afk Zool*, 20A:1, 1928.
Harris, H. A.: Endocranial form of gorilla skulls with special reference to the existence of dolichocephaly as a normal feature of certain primates. *Am J Phys Anthropol*, 9:157, 1926.
Hechst, B.: Ueber einen Fall von Mikrocephalie ohne geistigen Defekt. *Arch Psychiatr*, 97:64, 1932.
Holloway, R. L.: Cranial capacity of the Olduvai Bed I homine. *Nature* (Lond), 210:1108, 1966.

Holloway, R. L.: The evolution of the primate brain: some aspects of quantitative relations. *Brain Res*, 7:121, 1968a.
Holloway, R. L.: Review of man-apes or ape-men? *Hum Biol*, 40:421, 1968b.
Holloway, R. L.: Some questions on parameters of neural evolution in primates. *Ann NY Acad Sci*, 167:332, 1969.
Holloway, R. L.: New endocranial values of the australopithecines. *Nature* (Lond) 227:199, 1970.
Huizinga, J. and Slob, A.: Progressive brachycephalisation. *Proc Kon Ned akad Wet*, 68:297, 1965.
Hyden, H.: A molecular basis of neuroglia interaction, In Schmitt, F. O. (Ed.): *Macromolecular Specificity and Biological Memory*, Cambridge, MIT Pr, 1962, pp. 55-69.
Jerison, H. J.: Interpreting the evolution of the brain. *Hum Biol*, 35:263, 1963.
Kappers, C. U. A., Huber, G. C., and Cosby, E. C.: *The Comparative Anatomy of the Nervous System of Vertebrates, including Man*. New York, Hafner, 1936.
Keith, A.: *A New Theory of Human Evolution*. London, Watts, 1948.
Kumer, H.: Beitrag zur quantitativen Bestimmung der Entwicklungshohe des Saugetiergehirnes. *Psychiat Neurol* (Basel), 142:352, 1961.
Lancaster, J. B.: On the evolution of tool-using behaviour. *Am. Anthropol*, 70:56, 1948.
Lashley, K. S.: *Brain Mechanisms and Intelligence*. Chicago, U of Chicago Pr, 1929.
Lashley, K. S.: Persistent problems in the evolution of the mind. *Q Rev Biol*, 24:28, 1949.
Latimer, H. B.: The weights of the brain and its parts and the weight and length of the spinal cord in the adult male guinea pig. *J Comp Neurol*, 93:37, 1950.
Le May, M.: The language capability of Neanderthal man. *Am J Phys Anthropol*, 42:9, 1975.
Lenneberg, E. H.: *New Directions in the Study of Language*. Boston, MIT Pr, 1964.
Lilly, J. G.: *Man and Dolphin*. New York, Doubleday, 1961.
Mall, F. P.: On several anatomical characters of the human brain said to vary according to race and sex with special reference to the frontal lobes. *Am. J Anat*, 9:1, 1909.
Matiegka, H.: Uber das Hirngewicht, die Schadelkapazitat und die Kopfform, Sowie derm Beziehungen zur psychischen Thatigkeit des Manschen, *S B Kgl Bohmischen Ges Wiss Prag Math-nat Cl*, 20:1, 1902.
Mettler, F. A.: *Culture and the Structural Evolution of the Neural System*. New York, Amer Mus Nat Hist, 1955.
Noback, C. and Moscovitz, M.: Structural and functional correlates of

"encephalization" in the primate brain. *Ann NY Acad Sci, 102*:210, 1962.

Olivier, G. and Tissier, G.: Determination of cranial capacity in fossil man. *Am J Phys Anthropol 43*:353, 1975.

Oppenheim, S.: Zur Typologie des Primatecranium. *Z Morphol Anthropol, 14*:1, 1911-1912.

Pakkenberg, H. and Voigt, J.: Brain weight of the Danes. *Acta Anat, 56*: 297, 1964.

Pearl, R.: Biometrical studies in man. *Biometrika, 4*:13, 1905.

Pearl, R.: The weight of the Negro brain. *Science, 80*:431, 1934.

Pffister, H.: Die Kapazitat des Schadels beim Saugling und alteren Kinde. *Mschr Psychiatr Neurol*, 1903 (quoted by Zuckerman, S. 1928, q.v.).

Pietsch, P.: Shuffle brain. *Harper's Mag, 244*:41, 1972.

Pilbeam, D.: Early Hominidae and cranial capacity. *Nature* (Lond), *224*: 386, 1969.

Putman, C.: *Three New Letters on Science and Race.* New York, National Putman Letters Committee, 1963.

Putman, C.: *Race and Reality: A Search for Solutions.* Washington, Pub Aff Pr, 1967.

Radinsky, L.: Relative brain size: a new measure. *Science, NY, 155*:836, 1967.

Radinsky, L.: The fossil evidence of prosimian brain evolution. In Noback, C.(Ed.): *Advances in Primatology.* New York, Appleton, 1970.

Radinsky, L.: The fossil evidence of anthropoid brain evolution. *Am J Phys Anthropol, 41*:15, 1974.

Randall, F. E.: The skeletal and dental development and variability of the gori'la. *Hum Biol, 15*:236, 307, *16*:23, 1943-1944.

Rensch, B.: *Evolution above the Species Level.* New York, Columbia U Pr, 1960.

Schultz, A. H.: Growth and development of the chimpanzee. *Contrib Embryol Carneg Inst, 28*:1, 1940.

Schultz, A. H.: Observation on the growth, classification and evolutionary specialisation of gibbons and siamangs. *Hum Biol, 5*:212, 1933.

Schultz, A. H.: Past and present views of man's specializations. *Ir J Med Sci, 380*:341, 1957.

Schultz, A. H.: Die Schadelkapazitat mannlicher Gorillas und ihr Hochstwert. *Anthropol Anz, 25*:197, 1962.

Schultz, A. H.: The cranial capacity and the orbital volume of hominoids according to age and sex. In Homenaje a Juan Comas en su 65 Anniverario, Mexico, Libros de Mexico, 1965.

Selenka, E.: *Studien uber Entwickelungsgeschichte der Tiere.* Wiesbaden, Kreidel, 1899.

Shafer, E. A. and Symington, J.: *Quain's Elements of Anatomy.* London, Longmans, 1909.

Sholl, D. A.: The quantitative investigation of the vertebrate brain and the applicability of allometric formulae to its study. *Proc R Soc Med* (Lond), *135B*:243, 1948.

Spann, W.: Das Hirngewicht in Beziehung zur Todesursache under anderen Fakturen *Dtsch Z ges gerichtl Med, 44*:733, 1956.

Spann, W. and Dustmann, H. O.: Das menschliche Hirngewichte und seine Abhangigkeit von Lebensalter, Korperlange, Todeskeit von Lebensalter, Korperlange, Todesursache und Beruf. *Deutsche Z F Gerichtliche Medizin, 56*:299, 1965.

Straus, W. L.: *An Appraisal of Anthropology Today.* Chicago, U of Chicago Pr, 1953.

Stephan, H., Bauchet, R., and Andy, O. J.: Data on size of the brain and of various brain parts in insectivores and primates. In Noback, C. and Montagna, W.: (Eds.) *The Primate Brain: Advances in Primatology.* New York, Appleton, 1970.

Swan, D. A.: Juan Comas on "Scientific" racism again? *Mankind Monographs, 6*:24, 1964.

Symington, J.: Endocranial casts and brain form: a criticism of some recent speculations. *J Anat Physiol, 50*:111, 1916.

Tabulae Biologicae: Growth of Man, Vol. 20, 1941.

Thenius, E.: Stammesgeschichte der Saugetiere. *Handb d Zool Bd 8 (Mammalia)*, 47 Lieferung, 1-368.

Thoma, A.: Le caractere aromorphotique de l'evolution humaine as la lumiere des nouveaux fossiles. *Symp Biol Hung*, 9:39, 1969.

Tobias, P. V.: Brain size, grey matter and race—fact or fiction? *Am J Phys Anthropol, 32*:3, 1970.

Tobias, P. V.: *The Brain in Hominid Evolution.* New York, Columbia U Pr, 1971.

Todd, T. W.: *Growth and Development of the Child.* London, Century Co, 1933.

Tower, D. B.: Structural and functional organization of mammalian cerebral cortex: The correlation of neurone density with brain size. *J Comp Neurol, 101*:19, 1954.

Van Lawick-Goodall, J.: The behaviour of free-living chimpanzees in the Gombe Stream Reserve. *An Behav Mono, 1*:161, 1968.

Van Valen, L.: Brain size and intelligence in man. *Am J Phys. Anthropol, 40*:417, 1974.

Vallois, H. V.: La capacite cranienne chez les primates superieurs el le "Rubicon cerebral." *C R Acad Sci* (Paris), *238*:1349, 1954.

Washburn, S. L.: Tools and human evolution. *Sci Am, 203*:62, 1960.

Weidenreich, F.: Observation on the form and proportions of the endocranial casts of Sinanthropus pekinensis, other hominids and the great apes: a comparative study of brain size. *Palaeont sin Ser D 7 Fasc, 4*:1, 1936.

Weidenreich, F.: The skull of *Sinanthropus pekinensis*. *Palaeont Sin, 127*: 1, 1943.
Wilder, B. G.: Exhibition of, and preliminary note upon, a brain of about one half the average size from a white man of ordinary weight and intelligence. *J Nerv Ment Dis, 30*:95, 1911.
Witherspoon, Y. T.: Brain weight and behaviour. *Hum Biol, 32*:366, 1960.
Zuckerman, S.: Age changes in the chimpanzee, with special reference to growth of brain, eruption of teeth and estimation of age; with a note on the Taungs ape. *Proc Zool Soc* (London), *1*:1, 1928.

CHAPTER TWO

PRIMATE SKULL

INTRODUCTION

IN CONJUNCTION WITH the dentition, skull morphology has long been associated with taxonomy, although the cranial dimensions have been given varying weightings by different taxonomists (see Van Melsen (1972) for a philosophic discussion on the morphogenesis of the human head). Similar skull topographies, however, have evolved in widely divergent groups, indicating parallel or convergent evolution, perhaps due to the head organs themselves evolving along similar lines. for example, prosimians and simians show a tendency for a reduction in olfactory organs, development of color and stereoscopic vision, neocortical enlargement and specialization of the masticatory apparatus. All these skull changes occur concomitantly with phylogenetic changes of the head organs, however, and so are not significant alone for taxonomic discrimination.

In recent years, there has been much more progress in the understanding of primate skull structure (De Beer, 1937; Biegert, 1956; Moss and Young, 1960; Enlow, 1968; Moore and Lavelle, 1974). The skull is now regarded as a somewhat plastic structure which adapts itself to special mechanical and functional trends. These changes may be regarded as selectively determined adaptations to environment and behavior. This implies that the head organs can evolve independently of one another. For example, during the course of primate evolution, skull topography has been affected by enlargement of the neocortex and masticatory apparatus. The size of the masticatory apparatus also affects the "rolling up" of the neocranium and influences the degree of cranial base kyphosis, so that a masticatory apparatus large relative to the brain reduces the degree of "rolling up" (Moore and Lavelle, 1974) of the neurocranium. If two fossil forms are to be compared, therefore, estimation of the extent of neurocortical enlargement is possible from skull topography only if the relative sizes of the masticatory apparatus and brain are approximately equivalent. Hence, fossil australopithecine specimens can be compared with skulls of *Pan* and female *Gorilla*, but not with adult male *Gorilla*. Moreover, relative brain size to that of the masticatory apparatus seems to depend upon body size, with the masticatory apparatus being affected more than the brain by an overall increase in body size. From examination of the endocranial capacity and skull topography, australopithecines exhibit slight neocortical enlargement compared with pongids of the same body size. Thus, endocranial capacity, size of the masticatory apparatus or skull topography appear to have little taxonomic significance per se, without reference to body size coupled with the evaluation of other skull characters, which often is not possible in fossil specimens. Moreover, Giles (1956) remarks that allometric trends are very similar between *Pan* and *Gorilla*, suggesting that the "apparent difference in these categories (skull size) between chimpanzee and gorilla are due to similar ontogenetic growth patterns having been at some stage of evolution allowed to manifest different terminal overall morphological configurations through the mechanism of body volume increase."

The wide, flat, characteristically vertical face of man is a composite result produced by several morphogenetic circumstances associated with an enlarged cranium and orthograde posture. In man, the cranial floor is characterized by a distinctive downward bend with the foramen magnum located on its ventral side. This is related to an orthograde posture (Weidenreich, 1924) and marked expansion of the cranial fossae associated with cerebral expansion. Cerebral expansion produces a disproportionately enlarged cranial vault relative to the lesser growth of the cranial base. The kyphosis of the spheno-occipital complex provides an adaptive adjustment between these two differentially growing regions of the cranium and results in the ventral displacement of the foramen magnum compatible with the orthograde posture and the spheriodal configuration of the whole skull. These factors are associated with the formation of a distinctive "pocket" located anterior to the kyphosis of the cranial base beneath the overhanging prefrontal region of the cranium. This "pocket" is accompanied by the much changed facial complex. The human face, rather than comprising a projecting group of bones extending essentially forward from a horizontally orientated cranium, consists of an abbreviated composite of bony elements located in the space provided by the cranial flexure. In general, the face grows downward more than it moves forward, and this inferior manner of enlargement utilizes the region made available by the cranial flexure. Frontal and temporal expansion is directly associated with the distinctive forward-facing positioning of the two orbits and this arrangement also significantly reduces the extent of the nasal region located between the orbits and contributes to the vertically disposed nature of the nasal region.

SYNOPSIS OF PRIMATE SKULL FORM

In the tupaioid skull, the snout region comprises the maxilla and premaxilla separated by the premaxillary suture. The roof of the snout region comprises narrow elongated nasal bones, whereas the neurocranial roof and walls are formed by the parietal and occipital bones. The squamous part of the temporal bone furnishes the lower part of the lateral wall. Much of the cranial lateral wall supplies attachment for the temporalis muscles, demarcated superiorly by a temporal ridge, whereas posteriorly, this ridge runs into the transversely oriented nuchal crest. This nuchal area, and the occipital area posterior and inferior to it, provides attachment for the nuchal muscles. The orbital aperture is surrounded by a closed ring which is completed posteriorly by articulation of the orbital process of the frontal with the orbital process of the zygomatic bone. The auditory aperture is relatively large, rounded, and placed close to the posterior extremity of the zygomatic arch. Posteriorly are the occipital condyles for articulation with the atlas. The skulls of *Tupaia, Anathana,* and *Urogale* are very similar, except that the neurocranium is relatively more expanded associated with a more advanced cerebral development. The plane of the orbital apertures is directed more laterally and in the larger species, the snout region is extended more prominently.

Next to tupaioid skulls, lemurs retain the most primitive

mammalian features. The facial part of the skull tends to be large relative to the neurocranium, and the conspicuous and elongated snout is most strongly developed in the Lemuriforms (although in some of the more diminutive species of Lorisiforms, the snout is considerably reduced). The well-endowed brain is associated with a relatively voluminous neurocranium, although the skull lacks a conspicuous bony crest for attachment of the temporalis muscles and the foramen magnum is directed inferiorly and posteriorly. In all living lemurs, the orbital aperture is relatively large, especially in the more nocturnal species, e.g. galagos, pottos, and lorises, and the plane of the aperture tends to be directed inferiorly, although in lorises it appears superior. In fossil lemuroids, *Plesiadapis* shows primitive features of a small-sized neurocranium and the neurocranium of the *Adapidae* is not voluminous, with the result that there are strong sagittal and nuchal crests. The pronounced postorbital constrictions of the skull and small frontal bones reflect the relatively poor cerebral development and the foramen magnum is located posteriorly as in nonprimate pronograde mammals. In the Eocene types, *Smilodoctes,* the frontal part of the neurocranium is relatively more expanded and the sagittal crests largely obliterated due to the anterior separation of the temporal ridges. In some Eocene types, the orbital cavity is smaller than in modern lemurs. This suggests that either these creatures were not nocturnal or the visual apparatus was not highly elaborated. The morphology of Miocene *Progalago* is typically lorisiform, although the subsequent lemuroid evolution in Pleistocene gave rise to aberrant forms. In *Archaeolemur,* the general proportions of the skull are simian, with an abbreviated facial skeleton a large and globular neurocranium, anterior-facing orbital apertures and strongly developed nuchal crests. The general form of *Megaladapis* closely resembles large ungulates in that the neurocranium is relatively small with strong sagittal and nuchal crests, the facial skeleton is massive, the foramen magnum and occipital condyles face posteriorly, the zygomatic arch is broad and the orbital apertures are both small and surrounded by a raised bony rim.

Certain cranial characteristics of *Tarsius,* which resemble those found in higher primates, may be secondary to the enlarged orbital cavities. The reduction of the snout is at least partly a result of its posterior location, being overlapped by the orbits. The latter are surrounded by a bony ring and partly shut off from the temporal region of the skull by an incomplete bony wall, so approximating the anthropoid condition. The neurocranium is evenly rounded and there are no muscular crests. The foramen magnum is located inferiorly as in Anthropoidea, with a rounded occiput and a greater flexion of the basicranial axis than in lemurs. (Figs. 1, 2, 3, and 4)

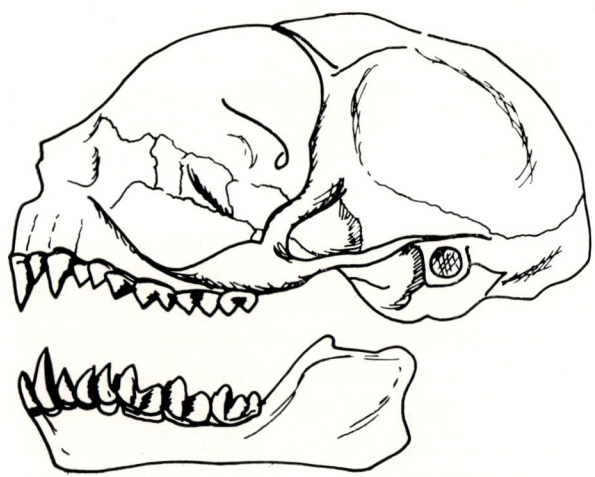

Figure 1. Diagram of skull of *Tarsius spectrum*.

Compared with lower primates, the neurocranium of monkeys is predominantly larger and dominates the small facial skeleton. This enlargement partly arises from expansion of the frontal and occipital cortical lobes and is mirrored by the extensive development of the frontal and occipital regions of the skull. The parietal bone forms a major part of the lateral wall of the neurocranium and articulates anteroinferiorly with an extension of the alisphenoid. This arrangement typifies New World and Colobinae of the Old World monkeys. In contrast, the squamous-temporal extends anteriorly to make sutural contact with the frontal bone, thereby separating the parietal from the alisphenoid in *Cercopithecinae*. The zygomatic bone is characteristically extensive and reaches

Primate Skull

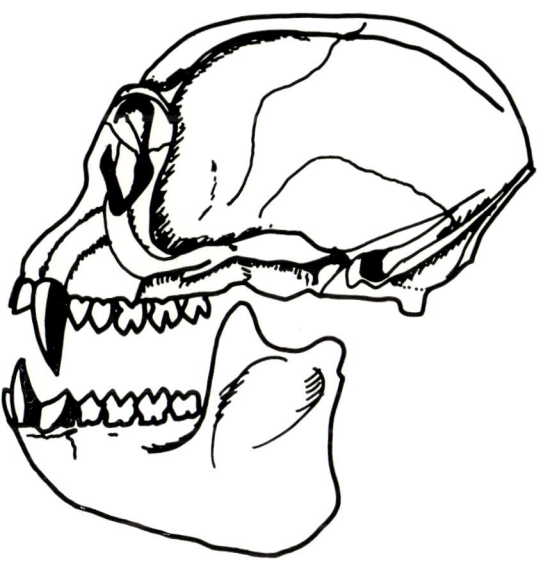

Figure 2. Diagram of skull of *Presbytis entel us*.

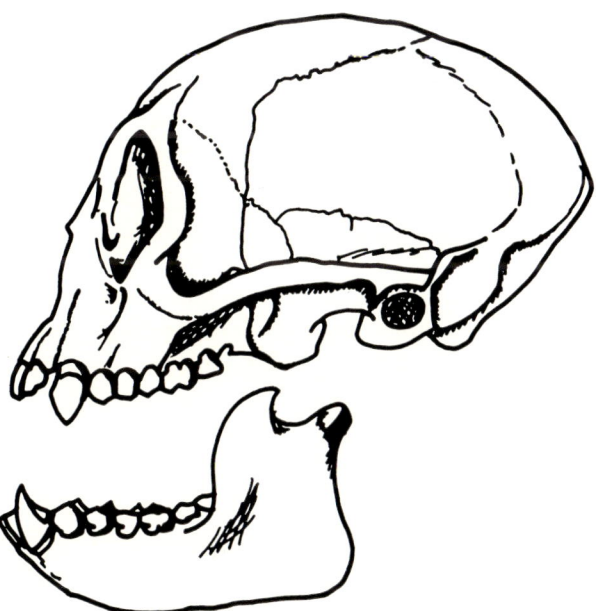

Figure 3. Diagram of skull of *Cebus fatuellus*.

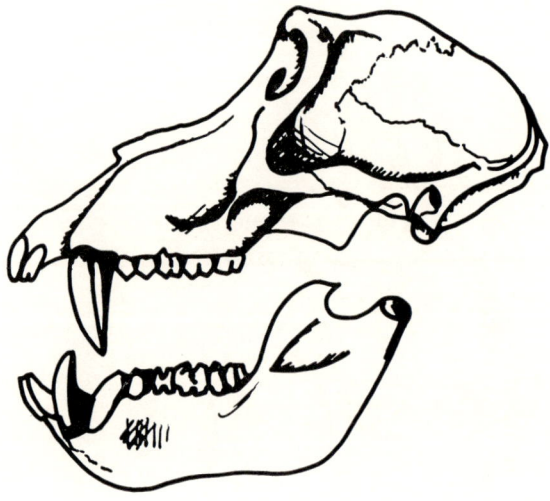

Figure 4. Diagram of skull of male *Papio anubis*.

the lateral skull wall to articulate with the parietal in platyrrhines, although this specialization is avoided on other anthropoids. The basicranial axis is flexed so that the facial skeleton lies inferior to the neurocranium in contrast to power primates, and similarly the foramen magnum is displaced more completely onto the basal aspect of the skull. The orbits are relatively large and directed anteriorly.

Due to the relative size of the neurocranium and facial skeleton, the skull of *Hylobatinae* exhibits superficial similarity to more primitive hominoids. Characteristically, the frontal bone extends far posteriorly onto the top of the skull, whilst the neurocranium is rounded and relatively smooth, although there may be a primordial sagittal crest. The occipital condyles are elongated and placed well posterior to the level of the auditory meatus. (Figs. 5, 6, 7, 8)

The skull of *Pan* and *Gorilla* share general morphological similarities. The orbital apertures are relatively large and rectangular, surmounted by the supraorbital torus. In the male *Gorilla*, there is a conspicuous sagittal crest associated with the relatively large jaws and this tends to coalesce with the nuchal crest posteriorly. The sagittal crest provides attachment for the temporalis

Primate Skull

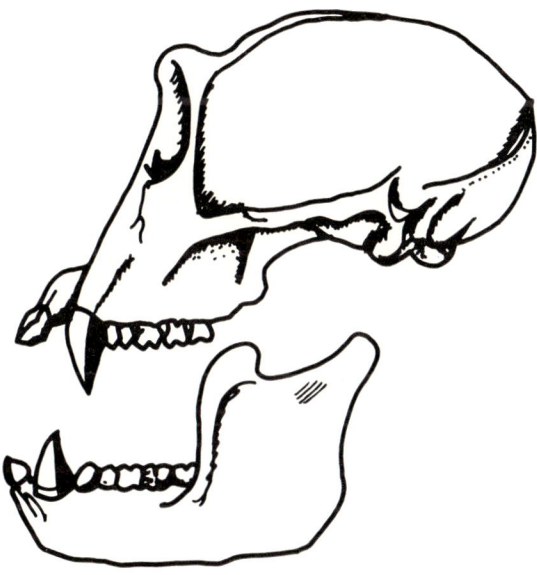

Figure 5. Diagram of skull of male *Pan troglodytes*.

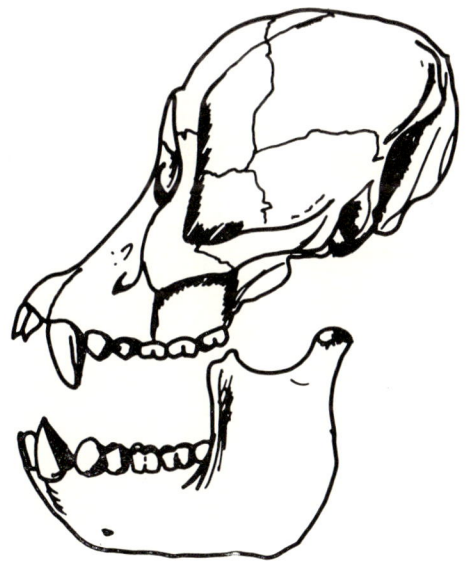

Figure 6. Diagram of skull of male *Pongo pygmaeus*.

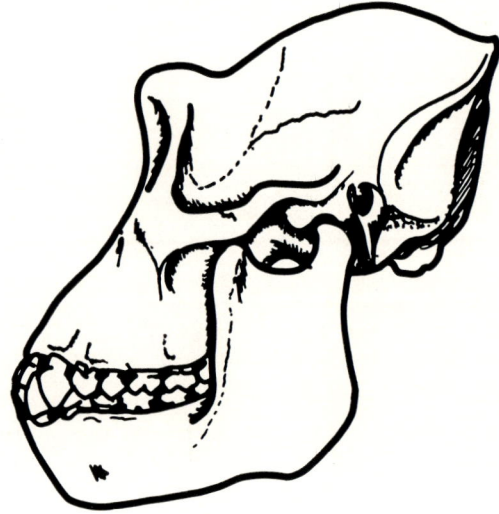

Figure 7. Diagram of skull of female *Gorilla gorilla*.

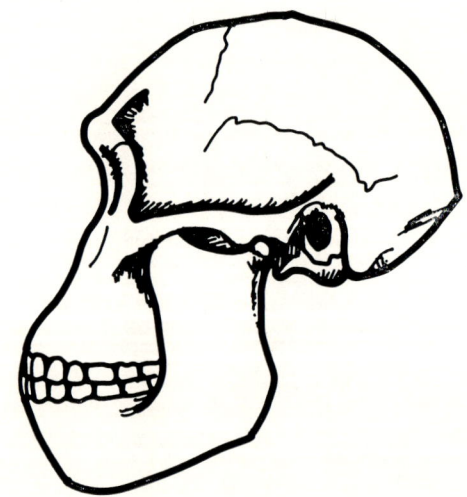

Figure 8. Diagram of skull of *Australopithecus*.

muscle and is rarely found in *Pan*. Nuchal crests characterize all anthropoid apes with their high location providing attachment for the nuchal muscles. The occipital condyles face postero-inferiorly, well posterior to the level of the auditory meatus. The skull of *Pongo* shows contrasts from *Pan* and *Gorilla* in that there is no continuous supraorbital torus, so that the frontal region appears more rounded. (In males the sagittal crest is as well developed as in *Gorilla*). The skulls of *Aegptopithecus* and *Dryopithecus* indicate that in Oligocene and Miocene, the pongid skull was more primitive than recent anthropoid apes, and in numerous features more closely resembles catarrhines than modern apes: small neurocranial size, narrow nasal apertures, less complete closure of the bony wall separating the orbital cavity from the temporal fossa, i.e. the cercopithecoid skull contrasts in morphology with the typically pongid dentition.

The distinguishing features of the skull of *Homo sapiens* are: a much larger neurocranium associated with a vertical forehead and rounded posteriorly directing occipital region, a reduced jaw and tooth size, a skull more evenly balanced on the vertebral column. Nevertheless, there are fossil Hominidae in which the skull lacks these strongly contrasted features or displays them incipiently. In fact, it is difficult to delineate sharp morphological boundaries between Pongidae either from Hominidae or Cercopithecidae on the basis of cranial features alone.

Although there are no sharp delineations between pongid and hominid skulls, various attempts have been made to quantify the distinguishing features, and this is not facilitated by the changes in endocranial capacity which occurred relatively late in hominoid evolution. This is particularly important in relation to skulls, such as *Australopithecus*, which demonstrates both hominid and pongid characteristics, although considered as a whole, the skull presents predominantly hominid features. These typically hominid features become greatly accentuated in more advanced types of the Hominidae. For instance, the neurocranium has already undergone considerable expansion by the beginning of the Middle Pleistocene, in the skull of *Homo erectus*, although still small by modern standards. This expansion is mainly associated with the occipital region, and partly accounts for the forward location of the occipital condyles relative to total skull length. The forehead region is still retreating and poorly developed, and the nuchal region is often strongly marked. Nevertheless, the skull of *Homo erectus* represents an intermediate stage between the anthropoid apes and *Homo sapiens*. By the end of the Middle Pleistocene, the hominid skull attained a degree of development very similar to modern man, with the exception of the strongly developed supraorbital ridges.

SIZE OF CRANIUM

Little work has been done on the relationship between nonmetric and metric cranial variation. Using American Negro males, Corruccini (1976) found that statistically significant associations between the total of fifty discrete and twenty-three metrical characters to be much more frequent than would be expected through random distribution. Multivariate analysis supplements simpler statistics by synthesizing patterns of variation within regions of the skull, identifying many interrelations of skull size and shape with discrete traits. A low but observable general influence is exerted upon nonmetric morphology by metrical variation of the human skull (or vice versa). Nevertheless, such important data have yet to be acquired for nonhuman primates.

In the skull of modern man, it is difficult to see how mechanical function might be related to variation except minimally. There are no muscles of any force other than those to the mandible, and these certainly are not adequate to account for much of the observed variation. Nevertheless, Hunt (1960) concluded that a hard diet increased the lateral growth of the palate and face of Austrian aboriginal children compared with Americans, and so dampened vertical growth of the face. Other special functional hypotheses have been addressed to the climate and nose shape (Weiner, 1954: Wolpoff, 1968), or intense cold adaptation and face shape (Coon, Garn, and Birdsel, 1950). These, of course, appeal to natural selection as the operative agent. Different and more general is the hypothesis that populations of major geographic zones are to be ranked as subspecies, i.e. their racial differences are to be traced to different genetic heritages from different hominid heritages, regardless of how the differences among the ancestors may have originally been established, rather than to a recent adaptive differentiation within the species (Coon, 1962).

The cephalic index lost relevance when skull shape was noted to be highly adaptive to the environment following the advent of modern genetic theory. Furthermore, in view of the multiplicity of bones participating differentially in the phenotypes, the statistical concept of the cephalic index cannot accurately reflect the genotype. The neurocranium tends to be spherical in form in order to provide maximum strength for a given weight and amount of material. Whereas the relative size of the neuro- and viscerocrania coupled with the nuchal musculature determine skull form, the brain is related to the endocranial capacity. Actually, the inner table of the neurocranium reflects brain size and form, whereas the outer table follows the structural demands of the other components of skull form (Moss and Young, 1960).

The overall effect of the masticatory apparatus on skull form is less than that of the brain, i.e. the jaws and face are merely minor appendages of the skull. This development has been quantified by Clark (1962), who computed the supraorbital region of the face. This index increases from the great apes, through *Australopithecus robustus* and *A. africanus* and so to *Homo erectus* and modern man, although the index has not been computed for other primate groups.

In the final stages of human evolution, the overall form of the neurocranium becomes increasingly spherical, less dolichocephalic and more brachycephalic. Brachycephalisation has been demonstrated in recent stages of human evolution, as well as in earlier times, and apparently reflects the continuing trend towards jaw and face reduction and improvement of cranial balance.

There are many theories to account for this brachycephalization. One holds that skull shape is affected by the bilateral compression exerted by the temporalis muscle, thereby preventing expansion of skull width. This thesis is not supported by morphological evidence of Reisenfeld (1955), who reported strongly developed temporal muscles in dolicho-, meso-, and brachycephalics. Nevertheless, Anthony (1903) and Hrdlicka (1924) have suggested that when mandibular size decreased during evolution, or as food became softer in recent times, the function of the temporalis was reduced, so that bilateral compression of the skull diminished resulting in a more brachycephalic skull form. Another theory holds that skull shape in its phylogenetic transformation is determined by increasing brain size in which the skull and face became shortened, the vault increased in both height and width above the cranial base and the occipital foramen moved anteriorly and changed in angulation so that a kyphosis developed in the skull base (Weidenreich, 1946). Another theory suggests that within a given population sample, there is a correlation between overall body height and skull shape, such that taller people become more dolichocephalic and shorter people more brachycephalic (Hooton, 1946). Allometric correlations of the skull, however, indicate the cranial form is more closely dependent upon facial length than other cranial parameters. For instance, a relatively large face is associated with a narrow relatively little-vaulted neurocranium and conversely a small face length is associated with a highly vaulted and widened braincase. Finally, Hooton (1946) considers that brachycephalisation results from hypothyroidism. Yet, despite all these theories, there is no corpus of data which favors one to the exclusion of others.

It is interesting to note that skull form influences facial form. For instance, dolicho- and brachycephalics are two extremes of skull form related to the more open (horizontal) and closed (upright) types of cranial floor flexure. These relationships in turn establish the basis for a series of other factors that have topographical results in the configuration of the face. The more closed type of cranial floor flexure associated with brachycephalics places the nasomaxillary complex in a more posterior and superior relative position in contrast to dolichocephalics where the face is located more anteroinferiorly. The mandible is aligned forward and upward in the first instance and downward and backward in the second. These features underlie the characteristic disposition among ethnic groups having a brachycephalic type of head for a more prognathic facial form, e.g. Japanese, whereas the caucasoid dolichocephalic groups have a tendency toward a more retrognathic facial configuration.

ORBITS

In primate evolution, the eyes and hence the orbits affect facial form. The basic structure of the eyes in higher primates is basically consistent (Woollard, 1927). Following the evolution of stereoscopic vision among primates, the eyes and hence their orbits, not only increased in size but also moved into a more central location on the face following a reduction in muzzle size. Also, with the reduction in muzzle size, the orbits retreated back into the head so that in man they lie closer to the anterior part of the brain than in other primates.

Schultz (1930) compared the orbits in 208 primates of varying ages and noted that the average size of the orbit is similar in man, chimpanzee, and orangutan, but smaller than in *Gorilla*. All large primates, however, possess proportionately smaller orbits than small primates, the relative capacity being greater in females than males, especially in the very young. The size of the orbit is essentially related to body size, although independent of genus, sex, and age. Only in some nocturnal primates have the orbits attained exceptional size, whereas the relative orbital capacity in recent man is the second lowest among all primates— the minimum size occurring in female *Pongo*—although the relative size is greater in fossil man.

The bony rim around the orbits may serve not only to protect the eyes (Boule, 1911), but also transmits forces developed by the masticatory apparatus (Weidenreich, 1946). The upper part of this bony rim—the supraorbital torus—assumes the greatest size in animals with a well-developed masticatory apparatus in which the eyes have not fully retreated under the forebrain, e.g. in *Gorilla* and *Pan*, whereas in man, *Pongo* and Hylobatinae, the torus is less well developed (Biegert, 1963). This torus is well developed in all the Hominidae, except in modern man, since it is only in the later stages of human evolution that the jaws finally retreated and the anterior region of the neurocranium expanded over the top of the eyes. The supraorbital torus is present in some monkeys and not in others, and is poorly developed in all immature skulls (Schultz, 1944). Moreover, the temporal crest often links up with the supraorbital torus, which suggests that the development of the temporalis muscle is associated with the supraorbital torus.

In higher primates, the optic axis (the axis of symmetry of the lens) and the bisual axis (the line through the center of the cornea to the fovea of the retina) approximately coincide, although in diurnal tree shrews, the optic and visual axes are almost at right angles to each other (Polyak, 1958). Even in man and the great apes, the axis of the conical eye socket diverges from the mid sagittal plane by approximately 23 degrees (Broca, 1873). Since the plane of the orbital margin is approximately perpendicular to the orbit axis in these giant primates, then neither visual nor optic orientation can be directly inferred from the orientation of the orbits.

Smaller animals must have relatively larger eyes to retain a degree of visual acuity comparable to that of larger relatives; therefore, relative eyeball diameter is inversely related to body size. This negative allometry influences convergence of the orbital margins. Other things being equal, orbital margin convergence shows a direct correlation with body size; an increase in relative diameter of the eye ball must be accomodated by an increase in transverse orbital diameter and this is most easily accomplished by caudal displacement of the posterior orbital margin along the zygomatic arch. This allometric change in orbital margin convergence does not necessarily reflect corresponding changes in the orientation of the optic axes.

NASAL FORM

The rhinarium was the most important sense organ in lower Primates, although with the ascendence of the visual sense the importance of olfaction is reduced in other primates. In higher Primates, both the size of the nasal cavity and complexity of the turbinate bones are very reduced and this appears to have occurred independently of muzzle and jaw reduction in modern man. Evidence indicates that the remote ancestors leading to the line of all higher mammals were small-long-snouted insectivorous mammals. Moreover, the faces of the surviving primates can be graded into two basic types: the fox-like lemur and the "quaint old man appearance" of catarrhines. In lower forms, e.g. lemurs, a rhinarium is present at the tip of a long snout, although following muzzle shortening, the rhinarium gave way to the true nose. This eventually developed so that it grew out between the nostrils.

In platyrrhines, the nostrils are widely separated and open laterally on each side of the broad median part of the nose. In contrast, the nostrils are drawn downwards and inwards towards the midline and so tend to make a V-shaped configuration in catarrhines, pongids, and hominids. A true snout is present in both lemurs and baboons and involves the maxillary and nasal bones as well as the vomer and mandible. Moreover, it is more clearly related to nasal than masticatory function, since long snouts are evident in animals without teeth. Apart from *Papio*, most of the higher Primates, especially the great apes, do not have a true snout. Rather in these latter, there is extensive development of the maxillary and mandibular alveolar processes, whereas the nasal bones are often greatly reduced. This alveolar growth results from massive dental development, whereas in *Papio* alveolar prognathism is superimposed on a well-developed snout. Furthermore, alveolar prognathism is less well marked in *Australopithecus* and similar but less extensive alveolar specialization is characteristic of the early forms of man although absent in recent times.

The human nose is relatively more prominent than that of most monkeys and apes. This is due partly to jaw recession and cerebral expansion which has resulted in only a small space remaining for the nasal cavity. Even in fetuses, there is wide variation in nasal form, with immature noses being characteristically wide and short with very low ridges: This infantile nose is retained in mongoloids. With development, the cartilaginous nasal septum grows forward and downward faster in man than in anthropoids and faster in caucasoids than negroids. Thus, in negroids, the everted lips and more protruding anterior teeth are associated with a shallower median septum and a reduced downgrowth of the nasal tip. The transverse components of growth are very marked in wide noses, broad nostrils and low brow ridges, e.g. australoids and negroids, and this tends to be associated with prognathous jaws and large teeth. Thus, a reduction in tooth size appears to have facilitated the vertical and forward nasal growth as in man (Manera and Subtelny, 1961: Posen, 1967; Kemble, 1973; Krogstad, 1973), whereas large teeth are associated with broad noses as in *Gorilla*. In addition, the ala cartilages appear feeble in *Pongo,* vigorous in *Gorilla,* and very vigorous in man. During evolutionary development, however, there seems to have been a general hominoid trend for inferior expansion of the nose as evidenced in Pongidae, Hominae, and Australopithecinae.

In modern man, the nasal index, which indicates the shape of the nasal openings, is highly dependent upon the humidity of the air (Weiner, 1954) so that many of the tropical races of man have flatter, more open noses than those inhabiting the drier areas. How far this is reflected in the evolution of the nose, however, requires more primate data.

THE BALANCE OF THE HEAD

The occipital condyles have shifted much nearer to the center of gravity of the human skull than any other primate. This is associated with the evolution of the erect posture and facilitates the balancing of the head on the vertebral column. Mechanically, the

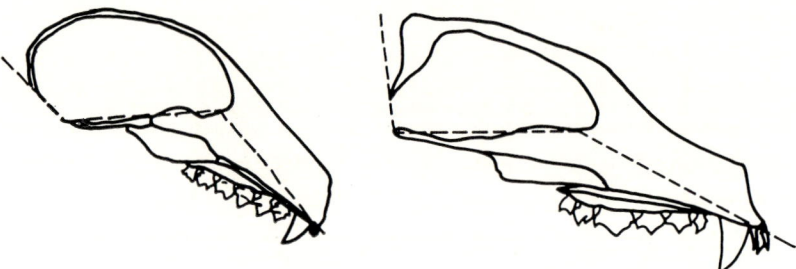

Figure 9. Sagittal skull section of lemur and dog, illustrating the relationship between the foramen magnum and basicranial axis.

skull represents a first-order lever, with the fulcrum at the center of the occipital condyles, the load comprising the weight in the precondylar portion of the head, and the power supplied by the nuchal muscles in addition to the postcondylar skull weight. Senyurek (1938) measured the position of the occipital condyles relative to the distance prosthion-opisthocranion in many primates. The results showed that the condyles lie aborally in howler monkeys and most orally in the squirrel monkey. In man, the relative position of the occipital condyles is constant and has

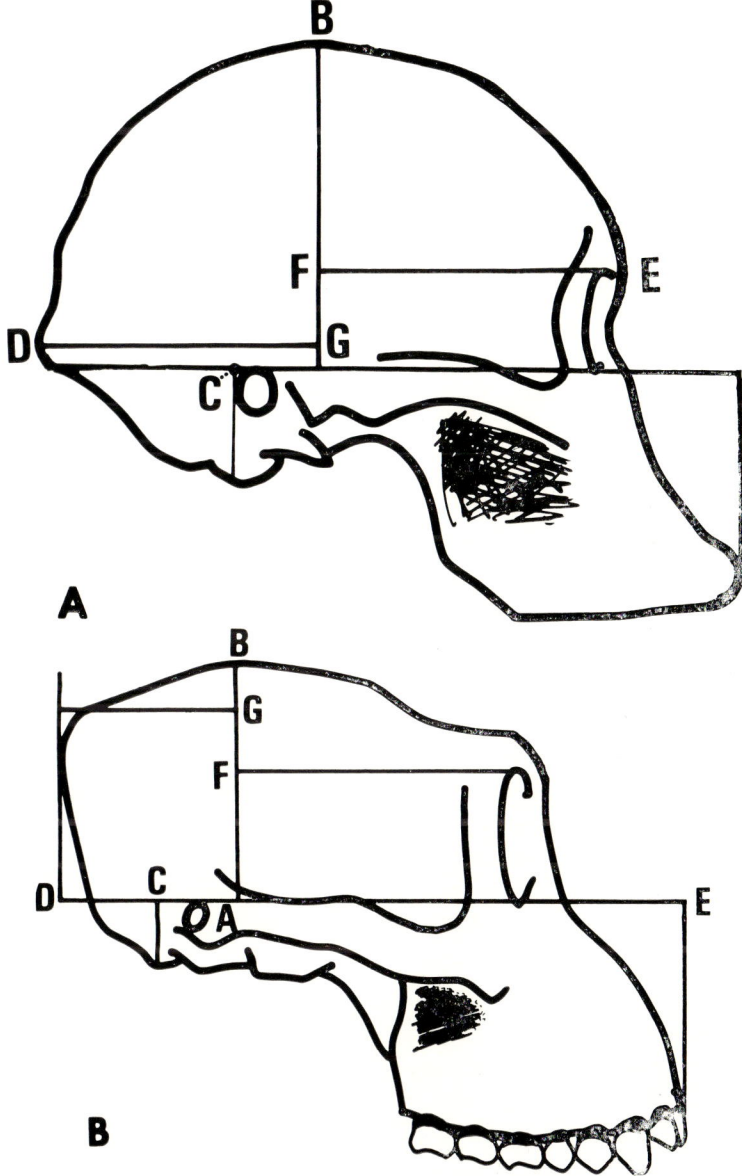

Figure 10. (A) Diagram of skull of male *Gorilla gorilla*
(B) Diagram of skull of *Australopithecus*
Cranial indices: Supraorbital height index $= \dfrac{FA}{AB}$:

Condylar position index $= \dfrac{CD}{CE}$:

Nuchal area height index $= \dfrac{AG}{AB}$.

shifted farther orally than in any other primate. Moreover, late in life, following the development of the occipital crest, the length of the postcondylar lever arm can be increased substantially, e.g. in male *Gorilla* and *Pongo*.

In 1940, Schultz constructed an instrument to determine the weights required to balance primate skulls when resting on the occipital condyles. The results showed that in order to lift the head at its most oral point requires an average force of 15.9 percent of head weight in adult man, much less than other adult primates which have a corresponding average of 37.3 percent. These values depend upon the weight of the precondylar portion of the head in excess of the weight of the postcondylar portion, and are affected by the distance between the fulcrum and the point at which the force is applied. If the head is balanced from the inion (the most aboral point from which the dorsal neck musculature acts) a force is required which in adult man averages 22.3 percent head weight with a range of 9.4 to 36.2 percent depending mainly upon the relative distance between the fulcrum and the location of the inion. In adult monkeys and apes, the corresponding values are much higher than the maximum for man, averaging 120.0 percent with the skull nearest to man being the squirrel monkey, *Saimiri*, and furthest away the howler monkey, *Alouatta*. These studies show, therefore, that less force is required to balance the human head than in adults of other primates. Also, the balance of the skull is very similar in young man and apes, although with increasing age, the human skull becomes more balanced and apes less balanced.

Dart (1925) suggested the notion of a head-balancing index which was used by Clark (1947) to affirm the forward position of the occipital condyles of australopithecine from Taung and Sterkfontein. Subsequently, Clark (1950) devised the condylar position index relating the central point on the convexity of the condyle to the maximum length of the cranium in the Frankfort Horizontal plane. The mean values (percent) of this index were 23.9 for adult male *Gorilla*, 23.0 for *Pan*, and 20.6 for *Pongo*. Ashton and Zuckerman (1951) computed a condylar position index, although the procedure was not identical to Clark's, in view of their measuring the pre- and postcondylar segments of the cranial length from the lowermost point of the condyle. Nevertheless, these workers confirmed that the location of the occipital condyles in australopithecines was not significantly different from gorillas, although significantly different from three series of modern human crania. Perhaps it would be preferable to compare *Australopithecus* with *Homo erectus*, rather than modern races of *Homo sapiens*, although such a study has yet to be undertaken.

The age changes of the condylar position index may be assumed to reflect the fact that from birth until all the permanent teeth have erupted, the face of apes grows relatively more than the cranium. Under such circumstances, the logical inference is that the load placed upon the nuchal musculature would be progressively increased. Such an increase in load would not automatically lead to a change in poise of the skull on the vertebral column. Indeed, if only the nuchal muscles were not equal to the additional forces now required of them, would it be expected that a change in the poise of the head would become observed, despite the increasing weight of the precondylar part of the head. This suggests that the nuchal musculature possesses sufficient responsive power to overcome the tendency for the head to sag forwards. The development of the nuchal crest during the time when the condylar position index is changing is now well established (Ashton and Zuckerman, 1956) and this may be taken as confirmation that, with a preponderance of precondylar growth the nuchal masculature increases its activity and its remodelling influence on the cranium.

The fact that the index for australopithecines is close to the upper range of male gorillas suggests that the skull is not as well balanced as in man. This conclusion is supported by the nuchal crests of australopithecines, which are stronger than in hominids. It might be that the nuchal muscles of australopithecines were more active in maintaining any sort of head balance against the twin challenges constituted by the situation of the condyles which are not as far forward relatively as in modern man, and by the excessive precondylar heaviness of the greatly elongated face and massive teeth. Confronted with these functional demands, it is possible that the head of australopithecines was not maintained as horizontally as in hominids, but sagged slightly forwards, though not to the same degree as in most of the Pongidae.

Rather than the location of the occipital condyles, it may be that the angle at which the condyles are set relative to the cranial base is related to the poise of the trunk. Moore, Adams, and Lavelle (1973) therefore studied the angle (the condylar angle) between the long axes of the occipital condyles and the vertical plane at right angles to the Frankfort Horizontal in mature and immature skulls of man, apes, and fossils. As with the location of the occipital condyles relative to the dorsal and ventral extremities of the skull (Ashton and Zuckerman, 1956), the results suggested a clear-cut contrast between, on the one hand, adult *Homo sapiens* and, on the other hand, extant apes all of which approximated relatively closely to one another. In the adult *Homo sapiens*, the high value of the condylar angle (close to a right angle) reflects the downward direction of the condyles, while in adult apes, the smaller size of the angle (centering around 60 degrees) reflects their more posterior direction. The condylar angle, however, provides an estimate of the orientation of the cervical vertebrae, since the cranial articular facets on the atlas vertebrae articulating with the occipital condyles are orientated in both man and apes at right angles to the general line of the cervical vertebrae. In addition, the values of the condylar angle were found to be similar in *Homo sapiens*, and in the broad category of Neanderthal man (LaFerassie, La Chapelle, and the Rhodesian skull). Furthermore, in immature *Pan* and *Gorilla* skulls, the size of the condylar angle approached human values, which is consistent with the view that such immature individuals move by brachiation (Ashton and Oxnard, 1964), i.e. generally have a more vertical orientation of their vertebral column than do adults of these genera. Nevertheless, the condylar angle need not necessarily be directly related to posture, as illustrated by the fact that although adult *Pongo* has an inclination of its occipital condyles not greatly different from that characteristic of adult *Pan* and *Gorilla*, its vertebral column, by a more frequent adoption by this species of a brachiating mode of progression (Ashton and Oxnard, 1964) probably approximates more closely to the vertical than either of the "knuckle-walking" African great apes.

Moreover, although the occipital condyles are angulated in australopithecines as in man rather than as in great apes, these findings alone need not be indicative of a habitual bipedalism. Such uncertainty derives from the unique combination of a human orientation of the occipital condyles with their ape-like positioning on the cranial base (Ashton and Zuckerman, 1951).

The position of the occipital condyles has been computed for only a few primate genera, and obviously more quantitative data are required before definite conclusions can be formulated. In addition, data have yet to be obtained as to whether the position of the occipital condyles is related to the location of the foramen magnum.

LOCATION OF THE FORAMEN MAGNUM

Dart (1925) devised a "head-balancing index" by expressing the basion-inion distance as a percentage of the basion-prosthion length. An index of 60.7 was computed for the Taung australopithecine which contrasted with 50.7 in an adult chimpanzee and 90.0 in dolichocephalic and 105.88 in brachycephalic Europeans. These varying indices were taken to indicate that australopithecines exhibited a more erect attitude than that of modern anthropoids. This index, however, would alter with other variables than the location of the foramen magnum. For instance, since the prosthion was regarded as the anterior terminal, varying degrees of proganthism would introduce additional variability. Other workers have, therefore, avoided such sources of error, by using the nasion or glabella as the anterior terminal, and projecting the position of the opisthocranion onto a baseline, either the Frankfort Horizontal or the nasion-opisthion base line. Indeed, from his studies of *Homo erectus*, Weidenreich (1943) concluded that in respect to the location of opisthion, the different stages of hominids show no recognisable tendency to approach the anthropoid more and modern man less. Hence, Weidenreich considered the more central the location of the foramen magnum of *Homo erectus* was in marked contrast to anthropoids.

While the angle subtended between the plane of the foramen magnum and the Frankfort Horizontal conveys an approximation of the orientation of the plane when the skull is held in a standard position, it provides no data on the relationship between the foramen magnum plane and the plane of the rest of the cranial base. But the bending of the cranial base will vitiate any conclusions so that the value of such a dimension is dubious. Hence, the taxonomic significance of the location of the foramen magnum has yet to be established.

MASTOID PROCESS

Due to the attachment of the sternomastoid muscle, the presence of the mastoid process is generally regarded as evidence of an orthograde poise.

Schultz (1950a, b) reported that in the large number of anthropoid crania examined, the mastoid region is extraordinarily variable. This workers noted that in *Pan* and *Gorilla,* true mastoid processes can appear late in growth, comparable in size and shape with those of man. In man, the mastoid process develops in all cases, whereas in apes they remain small except in a minority of specimens. In man, the mastoid processes begin to form early in postnatal life, but in apes they never appear noticeable before the approach of adulthood and then only in occasional specimens. Schultz concluded that the formation of mastoid processes is the result of an ontogenetic innovation which has led to the late development of these structures in only a part of the population of apes and to their comparatively early and constant development in man, so that in this restricted sense can the mastoid process be regarded as a distinction of man. Large mastoid processes of modern human shape are evident in mature australopithecines. In australopithecines, it would seem that these processes developed early in life and in both these respects, the Australopithecinae differ from the Pongidae and resemble modern Hominidae. Again, however, further data are required relating to other primate skulls before a taxonomic assignment can be attributed to the mastoid process.

QUANTITATIVE SKULL ANALYSIS

For a number of years, the morphology of the skull has been exploited in primate taxonomy. The morphology of the human skull has been described by a number of workers (Weidenreich, 1924; Morant, 1925; 1926; Kitson, 1931; Todd, 1932; Woo and Morant, 1934; Enlow, 1968). Other studies have provided metrical definitions of various primate skulls, e.g. *Pan* (Cameron, 1927; Keith, 1927; Krogman, 1931; Schultz, 1940; Fenart and Deblock, 1974), *Pongo* (Krogman, 1931; Schultz, 1941), *Gorilla* (Duckworth, 1895; Keith, 1927; Cameron, 1927; Schultz, 1927; Krogman, 1931), *Papio* (Goldblatt, 1926; Zuckerman, 1926) and *Macaca* (Moore, 1949) and fossil hominoids, e.g. *Australopithecus* (Broom and Schepers, 1946; Broom, Robinson and Schepers, 1950; Robinson, 1956; Dart, 1962; Tobias 1967), *Homo erectus* (Weidenreich, 1945; Arambourg, 1955; Howell, 1960), *Homo neanderthalensis* (Boule, 1911-1913; McGown and Keith, 1939) and *Homo rhodesiensis* Morant, 1928; 1930-1931; Weinert, 1936). A number of quantitative growth studies on the primate skull have also been reported (e.g. Krogman, 1931; Zuckerman, 1926; and Deniker, 1885), whereas numerous metric comparisons have been made between the skulls of different primate groups (e.g. Berry, and Robertson, 1910; Krogman, 1930; Enlow, 1966). As reviewed by Black (1931) and Weidenreich (1943), there are considerable difficulties in the measurement and definition of datum points in modern hominid and nonhominid primate skulls, and these difficulties are exacerbated when measuring incomplete fossil skulls. For instance, when measuring the cranial length in modern hominid crania, the inion and opisthocranion (furthest occipital point) are different points, the former low and the latter high on the cranium. In contrast, the two points coincide to mark the posterior termination of the maximum cranial length in many pongid, australopithecine, *Homo erectus,* and *Neanderthal* crania. In modern man, the glabella-opisthocranion length is essentially dependent on brain length. This contrasts with hominoids exhibiting a prominent glabella and with an opisthocranion thrown back onto the inionic summit of a large nuchal crest. Here the glabella-opisthocranion length depends upon cerebral length, degree of development of the supraorbital torus and the nuchal crest or occipital torus. Thus, the maximum cranial length reflects different biological parameters in modern man and in other extant or extinct hominoids.

64   *Evolutionary Changes to the Primate Skull and Dentition*

Similarly, the parameter of maximum cranial breadth presents difficulties. For instance, although determining the location of the euryon as the terminus of the greatest transverse breadth of the cranium presents no difficulty in recent man, this differs from the definition of maximum cranial breadth in *Homo erectus* and australopithecines.

Nevertheless, there are some metrical features, particularly relating to hominoids and pongids which are worth noting.

Tobias (1967) compared australopithecine parietal dimensions with those of samples of *Homo erectus* (from Java, China, and northwest Africa) and modern man. The values of the sagittal chord and arc of the australopithecines specimens were closest to *Homo erectus*, whereas the coronal and lambdoid marginal measurements were appreciably smaller than *Homo erectus*. In contrast, the ranges for these values overlap appreciably when *Homo erectus* is compared with modern man. Furthermore, based on the data of Martin (1928), the mean anteroposterior extension of the parietal bone in pongids is far smaller than in *Australopithecus*.

Parietal predominance over the occipital is essentially a feature of modern man (Sauter, 1941-1946), whereas *Homo erectus* exhibits the opposite condition. In this respect, *Australopithecus* is nearer to modern man than *Homo erectus*. Martin (1928) lists mean values for pongids, and these show a slight occipital predominance over the parietal, except in *Pan*, where there is occipito-parietal parity. Pongids are therefore in a neutral position of parieto-occipital equality, veering slightly toward occipital predominance; *Homo erectus* has clear-cut occipital precedence, whereas modern man, fossil sapient, European Neanderthals and australopithecines have clear-cut parietal preeminence.

The inion may be regarded as the dividing line in the median plane between the planum nuchale and planum occipitale of the occipital squama. Delattre and Fenart (1960) have shown how in ontogeny and growth of an anthropoid cranium with little cerebral expansion and with heavy ectocranial embellishments, the inion is carried progressively farther from its fetal or neonatal starting point. In modern man, with a much greater cerebral expansion and no marked migration up the calvarium of the nuchal muscles, an exactly opposite process occurs, with the upper (lambda-inion) segment of the squama increasing in modern man at the expense of the lower (opisthion-inion) component. Nevertheless, since *Homo erectus* on the one hand and *Australopithecus* and modern man on the other provide opposite extremes of this relationship, its taxonomic significance is difficult to appreciate: Additional primate data might clarify the situation.

In order to provide metrical data of the relationship between the neurocranium and the upper part of the face, Clark (1950) derived the supraorbital height index (percentage ratio between the height of the calotte above the Frankfort Horizontal). Ashton and Zuckerman (1951) computed the range of this index of 63 to 77 percent for three modern human cranial series and 49.2 to 54.0 percent in apes, with cranium V from Sterkfontein 74 percent. Again, however, more primate data are required relating to this parameter.

CRESTING

The pattern of cresting in hominoids has been the subject of much controversy, particularly since temporal crests may approach to within 7 mm of each other in Eskimos (Riesenfeld, 1955). Inspite of his lighter jaw and the larger insertion area for the temporal muscles, the temporal lines in (modern) man can move up so high as almost to meet in the median line. The appearance of a sagittal crest in australopithecines appears, therefore, to be only an extreme expression of this principle of variability (Riesenfeld, 1955). Generally, however, the sagittal crests in australopithecines differ from those in pongids, in that when present, they do not start from, nor extend back to, the external occipital protuberance.

Attempts have been made to correlate the thickness of the cranial wall during the course of human evolution with expansion of the neurocranium (Weidenreich, 1943). Both pongids and australopithecines are characterized by thin parietal bones as in modern man, so that it is the thickened parietals of *Homo erectus* which require explanation. Weidenreich (1941b) considered that the general massiveness (since thickening also occurs in the femur) is a primitive hominid character which disappears in the course of human evolution. Also, Darlington and Lisowski (1965) removed the temporalis muscle from young ferrets (*Mustela furo*) and noted that apart from a deflection of the sagittal crest toward the operated side, the bone in the operated side became thicker in the absence of the temporal bone than on the control side where the muscle was present. This suggests that the presence of the temporal muscle tends to check bone growth and keep the parietals thin, save for the crest at the spreading edge of the muscle.

There are two variables which systematically influence the relationship between temporalis muscle size and the area available for attachment. The first is the allometric relation between the size of the masticatory apparatus and that of the cranium. For example, Huxley (1932) noted in *Papio* that an increase of body size in the same or even very similar species leads to a far greater difference in the masticatory apparatus than in cranial size. This results in a disproportionate increase in the jaws and supporting musculature. The same relation holds for *Pan* and *Gorilla* where the summed posterior tooth areas for *Gorilla* is more than double that of *Pan*, while the average endocranial capacity is less than double. The effect of this allometric component is to reduce the relative area available for the attachment of the temporalis muscle with increasing body size. The second centers around individual variation resulting from both genetic and environmental influences with the latter especially acting on the size and robustness of the temporalis muscle. Furthermore, these variables are not mutually exclusive.

The pattern of cresting in *Australopithecus* is different from that of other hominoids (Wolpoff, 1974). Robinson (1958) pointed out that cresting in *Australopithecus* tended to be anterior compared with the pongids and because of the increased anterior component of the temporalis muscle, the temporal crest parallels the supraorbital margin to a far more medial position than is usually the case in pongids (Tobias, 1967). The development of the anterior component of the temporalis muscle (with the masseter muscle) possibly serves to maximize the power available to the posterior teeth and is possibly adaptive to a dietary regime including hard objects. Consequently, in larger australopithecines, teeth in close proximity to the anterior portion of the molar tooth row, e.g. the second permolar, tend to be molarized. Thus, the unique features in australopithecine crests are part of a morphological pattern adapted to generating powerful horizontal forces through the entire premolar-molar row during mastication. The occurrence of this morphological configuration suggests the presence of hard objects in the australopithecine diet.

STATISTICAL ANALYSIS OF SKULL DIMENSIONS

The use of nonmetrical variation of the human cranium, e.g. wormian bones (Bennett, 1965), torus palatinus (Dorrence, 1929; Wo, 1950) and torus mandibularis (Akabori, 1939; Johnson, Gorlin and Anderson, 1965) has increased considerably since Laughlin and Jorgensen (1956) demonstrated its potential for population studies, and a wide range of variants have been used by Berry and Berry (1967), Knip (1970), and Rightmire (1972) for the comparison of different population groups. The method depends on the premise that the incidence of the variants is genetically rather than environmentally controlled. In man, the precise contribution of the genotype is difficult to establish, although evidence suggests multifactorial control of variant incidence. Correlation studies in man (Hertzog, 1968) have shown that, in general, the variants occur independently of one another, which means that the use of a wide range of variants in population studies should be well fitted to demonstrate genetically determined affinities between samples, as illustrated by Berry (1974) on a survey of North European crania. Further studies on nonhuman primate groups are required however. Furthermore, Corruccini (1974) concludes that discrete traits in isolation are not of paramount value to skeletal genetic studies, although they may be vital in comparison and in conjunction with other types of data in analyzing the population genetics of extinct groups.

Subsequent to simple descriptions of a skull, any attempt to relate it to a population involves either a concious or unconcious analysis of variation and probability. The cruder the application of statistical procedures to the data, the less effective and more subjective is the analysis and interpretation. This crudeness may reflect either the paucity of the data, e.g. fragmented specimens, or the viewpoint of the investigator.

There are two basic problems. First, the determination of the variability of a population and, secondly, the placement of a single specimen with relation to a known population. At the turn of the century, Pearson and his coworkers employed univariate statistical analysis to compare skull populations. There are, however, limitations in the use of univariate analysis. If "n" dimensions are measured for a skull, a metrical profile is obtained. If, then, each of the "n" measurements is used by itself to find the population mean and standard deviation, the individual vector is dismembered, nor is a population vector created in the process. The statistics are simply those of a measurement vector, a vector of measurement "x" determined by "m" individuals. This describes one aspect of a population and permits it to be compared, e.g. by t-tests, with another population, more or less as one compares population in gross morphological features. But the real object of the study is the measurement concerned, not the populations, which simply provide the source of variation in the measurement. Thus, the limitation of univariate statistical analysis is that there is no real vector or profile representing either individuals or populations. In addition, population differences and relationships have to be inferred by a mental summation of the differences in separate measurements and their significance, without a suggestion as to which measurements are biologically significant and which are not. Closeness of mean measurements between population is regarded as indicative of relationship, yet small absolute differences which are in opposite directions in important features may provide quite a different picture as to shape. The one attempt to escape from such deficiencies has been the prolific use of indices, which actually do reflect some aspects of shape.

The use of univariate statistical analysis has the advantage that the results of analyses can be readily comprehended, a feature which cannot be stated, even at this time, for the results based on multivariate statistical techniques. Furthermore, if features are bold and obvious, then individual skulls may be classified by the experienced eye, i.e. statistical analysis of metrical profiles may be superfluous. Nevertheless, use of the experienced eye often merely reflects the reputation of the worker, rather than the objective significance of the skull in question.

In the 1950s, it was realized that any measurement, as a character, must have a context of other measurements of the same specimen, and in the same way, each specimen must have the context of populations similarly integrated as to their measurements or characteristics forming a "coherent matrix" as a background for interpretation. In multivariate analysis, therefore, the individual is not decomposed, but remains a vector of all his measurements taken together. Such a vector of measurements locates an individual as a point in space. This is multivariate space, which has as many dimensions as there are measurements, is created by the population as a whole, all individuals being located as points in the same space. Multivariate analyses consist in creating secondary or transformed variables on which to read individuals and populations, i.e. they take the form of reference axes in the space providing new sets of coordinates to locate the populations and individuals. Various multivariate techniques simply depend on the way the space is to be examined.

Only utilizing multivariate statistical techniques is it possible to examine the skull as a whole, yet even suitable statistical techniques are dependent upon adequate data. The more accurate the metrical profile, the more dimensions required, although in practice the number of dimensions measured is governed by expediency or by consideration as to which dimensions are biologically important: both decisions being subjective. Nevertheless, if different dimensions for two skull samples are compared by multivariate analysis, then it is possible that contrasting forms of discrimination between the samples may emerge from the analysis, so that their subsequent interpretation may be variable. Thus, apart from establishing the fact that multivariate analysis is the only means possible to compare skulls as biological entities (Landauer, 1962; Brown, Barrett and Darroch, 1965), the method of acquiring metrical data requires a great deal more research.

Nevertheless, Howells (1972) applied factor and discriminant analysis to the same body of cranial measurements of modern man and indicated that interpopulation differences in recent man involve the same morphological patterns as does individual variation within the population. These patterns do not follow such classic distinction as cranial or nasal incides. Rather, they consist of general differences in facial height, vault and base breadth, forward prominence of lateral borders of the face, upper facial broadening, etc., as well as more anatomically local differences in such features as horizontal profile of the orbital margin of the frontal bone, midfacial flatness, occipital sagittal contour, etc., to a total of not less than eighteen factors from the measurements used. Variation in total cranial length and breadth of the nasal aperture are notable for their lack of importance in these cranial patterns. Obviously, this type of research must be applied to other primate skulls.

VARIATION IN CRANIAL FORM

As in other parts of the skeleton, there are two sources of variation in skull form—genetic and nongenetic—although the relevance of each factor remains conjectural. Evidence has been accumulating in favor of the notion of Brodie (1941) that skull, face, and dental development is based on an inheritable growth pattern, the genetic factors involved acting separately on different components of the craniofacial complex. Nevertheless, the identity of these components and their specific modes of genetic control remains largely unknown.

One of the major problems which has delayed investigation into the influence of heredity is the complexity of multifactorial inheritance. Virtually all dentofacial characteristics are polygenic and continuously variable and traditional types of cephalometric analyses involving line or angle constructs or dimensions provide only limited data for heredity studies. These constructs are generally defined in terms of widely separated datum points, often spanning several anatomic structures and growth sites. Also, each dimension may be influenced by changes in any or all of the structures. Hence, only a cautious interpretation can be extrapolated from such data.

Several investigators have suggested that the morphological aspects of single bones or bone segments, as expressed by their contour traced from cephalographs may be the best indicators of genetic control in the craniofacial complex. Indeed, Kraus, Wise, and Frei (1959) suggested that the contour of a bone segment is an expression of the "total morphologic configuration" of that bone and so may reflect genetic control of the bone. More recently, Harris, Kowalski, and Watnick (1973) have compared the similarity of nine small bony contours, six in the mandible and three in the maxilla, in groups of related and unrelated individuals. The data provided evidence of strong familial patterns for most of the coutours considered, although they did not provide much of an insight into specific modes of inheritance. Much more data are therefore required before the effects of genetic and nongenetic factors on craniofacial form can be established.

The marked variation in both neuro- and viscerocrania of man contrasts with the apparent homogeneity of catarrhines.

The comparative homogeneity of catarrhines is evident from examination of the neurocranium. For instance, the allometric changes in size of the neurocranial cavity when related to body weight fit closely on a single curve for all Old World monkeys, whereas corresponding data for platyrrhines follow separate curves for marmosets, howler monkeys, and spider monkeys respectively, and the three families of hominoids also differ widely in their relative endocranial capacities (Schultz, 1956).

The relative location of the occipital condyle on the skull base changes ontogenetically in varying degrees in all simian primates so that it can differ strikingly among adults. This character can be expressed by the percentage relationship between the distance from the nasion to the center of the occipital condyle and the nasion-opisthocranion length. This is practically 100 percent in all mammals, but varies between 65 to 75 percent at the time of birth and changes in adults to average 76 percent in *Saimiri* to 96 percent in *Allouatta* among the platyrrhines, from 66 percent to 95 percent in *Pongo* among the hominoids, and 87 percent in *Macaca* to 93 percent in most of colobines among the cercepithecoids (Schultz 1955). The remarkable uniformity of the entire group of Old World monkeys is illustrated by the universal absence of the frontal sinuses, which contrasts to the development of such in some but not all of prosimians, platyrrhines, and hominoids (Weinert 1926). The craniofacial skeleton has generally acquired much more marked and significant taxonomic differences compared with the neurocranium. It not only changes more profoundly in the postnatal period, but also varies individually more intensively (Schultz 1962, Vogel 1966).

The endless variation in size, position, form, and proportions of the structures comprising the dentofacial complex makes it difficult to discriminate between important and secondary factors affecting facial form. These factors include genetic (Trevor, 1947; Bjork, 1972), litter size (Park and Nowosielski-Slepowron, 1972), asymmetry (Woo, 1931), old age (Tallgren, 1957; Brodie, 1953), body build (Bjork, 1955), cold (Steegman, 1970; Beals, 1972), nose shape (Glanville, 1969), mouth shape (Yamazaki, 1969), arch shape (Hasund and Sivertsen, 1971), tooth shape (Yordanov, 1969), tooth position (Wylie, 1955), and tooth loss (Michelis, 1963). In view of this variability, many attempts have been made to categorize craniofacial morphology (e.g. Stockard, 1923), although not only has the definition of these categories yet to achieve universal acceptance, but also such a system has yet to be formulated to include nonhuman primate groups, and a proportion of this variability is contributed by the mutual structural influence of the various elements of the face upon one another.

This is illustrated when asymmetry is considered in primate skulls. Woo (1931) carried out direct chordal and arcual measurements of a large number of human skulls from the Twenty-sixth to Thirtieth Egyptian Dynasties and reported asymmetry of the right side compared with the left, reflecting the development of the right cerebral hemisphere. The contralateral side of the facial complex exhibited asymmetry with the left zygoma and left maxilla being larger. Groves and Humphrey (1973) reported an asymmetry of *Gorilla* with the left side exhibiting a marked increase in length of the temporal fossa to the gnathion. These workers postulated that such asymmetry may result from an asymmetry of function of the masticatory system. The skull complex comprises numerous constituent parts, so the degree of harmony between the parts determines the symmetry of the whole. Vig and Hewitt (1974) examined sixty posteroanterior cephalometric radiographs of "normal" children, who clinically exhibited facial symmetry. An overall asymmetry was noted in most cases, with the larger side to the left. The cranial base region, lower maxillary region and mandibular region showed a left-sided excess, whereas rightsided excess was apparent for the maxillary region. These findings suggest a compensatory adaptation during growth to effect an integration of the facial components, and the mechanism whereby this occurs may be part of an evolutionary process.

It is reasonable to assume that optimal function is provided by maximum cuspal interdigitation of teeth and this relationship is possible even though facial asymmetry may still exist. If this were not the case, then an asymmetrical functional activity of both temporomandibular joint mechanisms must compensate during masticatory and nonmasticatory activities in which the teeth are approximated. In clinical practice, this is frequently associated with pain and dysfunction and so is not a normal adaptation in man. Hence, to enable bilateral symmetrical function and maximum intercuspation of the teeth to occur, compensatory changes seem to be operating in many during growth and development of the dentoalveolar structures which minimizes the underlying asymmetry in the spatial arrangement and size of the jaws. This factor is no longer essential for man's survival with his modern diet. Nevertheless, this is a possible factor in the evolution and natural election processes of other primates.

Another source of variability is craniofacial growth which has been examined in apes and man (Krogman, 1930), in *Gorilla* and *Pan* (Krogman, 1931), in *Pan* (Gavan, 1953), *Cercopithecus* (Preston and Evans, 1976) and in *Macaca* (Moore, 1949; Enlow, 1966; Gans and Sarnat, 1951). No further description is required here, except to note that some variation in craniofacial dimensions between one primate group and another, or between two different population samples, may merely reflect variation in craniofacial growth. Also, although apparently unaffected by cerebral growth (Baer and Harris, 1969), the craniofacial growth pattern may be affected by a number of environmental factors, including: growth at the midpalatal suture (Heflin, 1970), growth of the nasal system (Young, 1960), tooth eruption, size, (tooth length and occlusion (Riesenfeld and Siegel, 1970), and muscles of mastication (Moller, 1966).

Many investigators have described changes in the size of the component parts of the face with age, and it is generally held that the increase in their linear dimensions does not occur at a uniform rate, but proceeds with alternate acceleration and deceleration (Davenport, 1940; Thurow, 1961). Brodie (1941), however, considered that the rate of growth diminished progressively at least up to eight years of age in man. Subsequently, Gilda (1974) has found that the shape of the distance-travelled curve for eleven facial dimensions in one subject over a period of eighteen years varied from a resemblance to certain body length curves in boys (sella-gonion and nasio-menton) to a similarity with the neural curve of Scammon (nasion-sella). The other dimensions described a character of their own which could not be categorized according to any one of the classical types of tissue growth curves. The growth curves were generally parabolic in form and seemed to represent the portion of the S-shaped growth curve which occurred after the major inflection point. The instantaneous relative growth rates of the dimensions studied were constant over certain sections of the curves, changed abruptly and decreased progressively with age. Indeed, a decreasing facial growth gradient was apparent, but it described a stepwise rather than a continuous pattern. Much more data are necessary before similar growth patterns can be identified for other nonhuman primates.

CRANIOFACIAL CORRELATIONS

Woo (1931) examined 800 male crania previously used by Pearson and Davin (1924), and on each specimen sixty-three dimensions were measured; fifty of these were corresponding measurements of homologous bones and each related to a single bone. Woo found that most of the homologous bilateral measurements were strongly correlated. On the same series of skulls, Pearson and Woo (1935) found that identical indices of homologous bones were closely correlated, that different indices on the same and on homologous bones were less strongly associated, and that the correlation between indices on different bones on the same or opposite sides of the cranium were all very low, as were the correlations between linear measurements on various contiguous bones. Thus, of the factors whose influence on the correlations was examined, homology was by far the most powerful.

Subsequently, Lindegard (1953) published a large number of correlations between twenty-three and thirteen angular profile cephalometric measurements based on 243 subjects and noted the presence of an association between cranial base flexure and prognathism, and between facial heights and angles between the horizontal facial planes with the shape of the mandible. Many cephalometric correlation studies have since appeared and these have been mainly characterized by the interpretations of individual associations.

By far the most significant study to date is that of Solow (1966), who studied 102 males aged twenty to thirty years. Correlation analysis showed that the body measurements, e.g. stature, were positively associated with most of the linear measurements and factor analysis revealed that the associations between extremity lengths and cephalometric measurements were evident in the form of loadings of measurements of jaw size on the length factor. The associations between extremity widths and cephalometric measurements were evident indirectly in the form of loadings of extremity widths on the length factor, and directly in the form of loadings of extremity widths on several cephalometric factors.

A large proportion of the correlations between the cephalometric variables were significant and often very big, although no association between cephalometric length and width measurements were detected. Fairly strong, positive correlations were found between maxillary width, anterior and posterior maxillary height and clivus length, and negative correlations between all these dimensions and cranial base flexure. That maxillary width and posterior height were negatively correlated with cranial base flexure indicates that a bent cranial base is associated with a wide and high maxilla. This suggests the presence of a compensatory mechanism that tends to keep the nasopharyngeal volume independent of cranial base flexion.

Positive association of jaw measurements with the length and width of the dental arches and tooth size were reported, and the length and prognathism of the maxilla were found to be associated with tooth size and dental arch length, although no corresponding associations were exhibited by the length and prognathism of the mandible. The correlations with tooth size were higher than those with the dental arch length. This suggests that there is an association between tooth size in both jaws and the basal length of the maxilla and that this may be responsible for the association between length and prognathism of the maxilla and the dental arch length in both jaws.

Face and jaw widths were found to be correlated with dental arch width in both jaws, with the widths of the maxillary and mandibular arches being more strongly associated with maxillary width than with other face and jaw widths. Factor analysis showed that this was due to a general association between dental arch and face width and a specific association between the width of the dental arches and maxillary width.

The length and prognathism of the mandible were positively associated with maxillary incisor inclination, maxillary arch width and transverse inclination of the maxillary lateral segments, and negatively associated with mandibular incisor inclination and mandibular alveolar prognathism. An increase in mandibular length and prognathism is thus correlated in the dentition by proclination of the maxillary incisors, retroclination of the mandibular incisors, lateral tipping of the maxillary lateral segments and an increase in maxillary arch width.

No association was detected between the dental arch widths and sagittal jaw relationship, although in the transverse plane there was an association between jaw width and transverse inclination of the lateral segments of the opposite arch. Correlation analysis also disclosed an association between maxillary width and transverse inclination of the mandibular arch, and between mandibular width and transverse inclination of the maxillary arch. Thus, just as deviations in sagittal jaw relations were associated with dentoalveolar changes in the transverse plane, so were deviations in the transverse jaw relation associated with dentoalveolar changes in the sagittal plane.

Finally, mandibular length and prognathism were associated with sagittal occlusion measurements and overbite was negatively associated with face height measurements, mandibular inclination, and mandibular length. Factor analysis showed that these associations were due to an association between the occlusion as a whole and the length and prognathism of the mandible and a negative association between overbite and mandibular inclination.

Although this study of Solow (1966) is the most comprehensive to date, it was based on a heterogeneous human population sample. It is not possible, therefore, to ascertain whether these relationships are also pertinent to other human population samples or to other primate groups. Nevertheless, it serves to emphasize that in order to obtain any meaningful information, the skull must be considered as a biological entity.

Indeed, craniofacial measurements have often been treated as independent variables in studies of heredity, despite the fact that various aspects of the craniofacial skeleton are obviously interrelated. Since it is possible that genetic and environmental factors influence multiple craniofacial measurements in a complex manner, an analysis of the correlations among the measurements would seem to be a more logical and reasonable approach to understanding the inheritance of these interrelated characters. Indeed, in order to understand the genetic control of craniofacial morphology, it may be more appropriate in future to examine the relevant variables as multivariate factors rather than to treat them as independent traits. Certainly if this approach were applied to various primate groups, then a considerable advancement in understanding of the craniofacial morphology would be achieved. This is illustrated by the fact that when multivariate statistical analysis is applied to the mandibular and tooth dimensions of *Gigantopithecus*, this group of fossils appears an aberrant form, less related to australopithecines and gorillas, than the latter are to each other (Corruccini, 1975). In another study based on 104 female and 188 male caucasoid students, Harris and Kowalski (1974) reported a poor correlation between tooth size and body size, although Ingervall and Lennartsson (1972) noted that the facial dimensions were highly correlated with the dental arch dimensions. Furthermore, Ingervall and Lewin (1974) describe a general positive corelation in size between the cranial base, neurocranium and viscerocranium.

CONCLUSION

The orientation of the facial skeleton, including the dentition, to the anterior cranial base is characterized by a wide range of variability, as judged from angular and linear measurements of cephalographs on a wide variety of primates. In contrast, the intrafacial orientation of the occlusal aspect of the dentition can be defined within narrow parametric estimates. These angular and linear orientation characteristics of small variability indicate a high order of morphologic integration between the facial skeleton and the dentition. The resulting pattern commonality reflects the integration of numerous growth and neuromuscular control mechanisms which function to attain and maintain biologically normal occlusal relationships. Indeed, the uniquely invariant nature of these occlusofacial relationships and their intimate correlation with function provide a frame of reference whereby normal and abnormal growth, development, form and function can be critically evaluated in a total facial context. Indeed, only by consideration of the skull or facial skeleton as a whole is comprehension of its complexity likely to ensue.

REFERENCES

Akabori, E.: Torus mandibularis. *J Shanghai Sci Inst,* 4:239, 1939.

Anthony, R.: L'evolution du pied humain. *Rev Sci Paris,* 1903:129, 1903.

Arambourg, C.: A recent discovery in human palaeontology, Atlanthropus of Ternifine. *Am. J Phys Anthropol,* 13:191, 1955.

Ashton, E. H. and Oxnard, C. E.: Locomotor patterns in primates. *Proc Zool Soc* (Lond), 142:1, 1964.

Ashton, E. H. and Zuckerman, S.: Some cranial indices of *Plesianthropus* and other primates. *Am. J Phys Anthropol,* 9:283, 1951.

Ashton, E. H. and Zuckerman, S.: Cranial crests in the anthropoides. *Proc Zool Soc* (Lond), 126:581, 1956.

Baer, M. J. and Harris, J. E.: A commentary on the growth of the human brain and skull. *Am J Phys Anthropol,* 30:39, 1969.

Beals, K. L.: Head form and climatic stress. *Am J Phys Anthropol* 37:85, 1972.

Bennett, K. A.: The etiology and genetics of wormian bones. *Am J Phys Anthropol,* 23:255, 1965.

Berry, A. C.: The use of non-metrical variation in the crania in the study of Scandinavian population movements. *Am J Phys Anthropol,* 40:345, 1974.

Berry, A. C. and Berry, R. J.: Epigenetic variation in the human cranium. *J Anat, 101*:361, 1971.

Berry, R. J. A. and Robertson, A. W. D.: The place in nature of the Tasmanian aboriginal as deduced from a study of his calvarium. *Proc Ro Soc* (Edin), *31*:41, 1910.

Biegert, J.: Das Kiefergelenk der Primaten, seine Altersveranderungen und Spezialisationen in Gestaltung und Lage. *Morph Jb*, 97:249, 1956.

Biegert, J.: The evaluation of characteristics of the skull, hands and feet for primate taxonomy. In Washburn, S.L., (Ed.): *Classification and Human Evolution*. Chicago, Viking Fund, 1963, pp. 116-144.

Black, D: On an adolescent skull of *Sinathropus pekinensis* in comparison with an adult skull of the same species and with other hominid skulls, recent and fossil. *Palaeontal Sin* 7:1, 1931.

Bjork, A.: Cranial base development. *Am J Orthod, 41*:198, 1955.

Bjork, A.: The role of genetic and local environmental factors in normal and abnormal morphogenesis. *Acta Morph Neerl Scand, 10*:49, 1972.

Boule, M. L'homme fossile de La Chapelle-aux-Saints. *Annle de Paleont,* 6,7,8, 1911-1913.

Broca, P.: Quelques resultats de la determination trigonometrique de l'angle alveolo-condylien et de l'angle biorbitaire. *Bull Soc Anthrop Paris ser 2,8*:150, 1873.

Brodie, A. G.: On the growth pattern of the human head. *Am J Anat, 68*: 209, 1941.

Brodie, A. G.: Late growth changes in the human face. *Angle Orthod, 23*: 146, 1953.

Broom, R. and Schepers, G. W. M.: The South African fossil ape-men, the Australopithecinae *Transv Mus Mem*, No. 2, 1946.

Broom, R., Robinson, J. T., and Schepers, G. W. H.: Sterkfontein ape-man, *Plesianthropus*. *Transv Mus Mem*, No. 4, 1950.

Brown, T., Barrett, M. J., and Darroch, J. N.: Craniofacial factors in two ethnic groups. *Growth, 29*:109, 1965.

Cameron, J.: The main angle of cranial flexion. *Am J Phys Anthropol, 10*: 275, 280, 286, 1927.

Campbell, B.: Quantitative taxonomy and human evolution. In Washburn, S.L. (Ed.): *Classification and Human Evolution*. Chicago, Viking Fund, 1963, pp. 50-74.

Clark, W. E. Le Gros.: Observations on the anatomy of the fossil Australopithecinae. *J Anat* (Lond), *81*:300, 1947.

Clark, W. E. Le Gros.: New palaeontological evidence bearing on the evolution of the Hominoidea. *Q J Geol Soc* (Lond), *105*:225, 1950.

Clarke, W. E. Le Gros.: *The Antecedents of Man*. Edinburgh U Pr, 1962.

Coon, C. S.: *The Origin of Races*. New York, Knopf, 1962.

Coon, C. S., Garn, S. M. and Birdsell, J. B.: *Races: A Study of the Problems of Race Formation in Man*. Springfield, Thomas, 1950.

Corruccini, R. S.: An examination of the meaning of cranial discrete traits for human skeletal biological studies. *Am J Phys Anthropol, 40*:425, 1974.

Corruccini, R. S.: Multivariate analysis of Gigantopithecus mandibles. *Am J Phys Anthropol, 42*:167, 1975.

Corruccini, R. S.: The interaction between non metric and metric cranial variation. *Am J Phys Anthropol, 44*:285, 1976.

Darlington, D. and Lisowski, F. P.: Changes in the sagittal and nuchal crests of the skull of the ferret after partial removal of the temporal muscles. *VIII Int Cong Anat, 1965.*

Dart, R. A.: *Australopithecus africanus:* the man-ape of South Africa. *Nature* (Lond), *115*:195, 1925.

Dart, R. A.: A cleft adult mandible and nine other lower jaw fragments from Makapansgat. *Am J Phys Anthropol, 20*:267, 1962.

De Beer, G. R.: *The Development of the Vertebrate Skull.* London and New York, Oxford U Pr, 1937.

Delattre, A. and Fenart, R.: *L'Hominisation du Crane.* Paris, Centre National de la Recherche Scientifique, 1960.

Deniker, J.: Le developpement du crane chez le gorilla. *Bull Soc Anthropol* (Paris), 8:703, 1895.

Dorrance, G. M.: Torus palatinus. *Dent Cosmos, 71*:275, 1929.

Duckworth, W. L. H.: Variations in crania of Gorilla savagei. *J Anat, 29*: 335, 1895.

Enlow, D. H.: A morphogenetic analysis of facial growth. *Am J Orthod, 52*:283, 1966.

Enlow, D. H.: *The Human Face.* New York, Har-Row, 1968.

Fenart, R. and Deblock, R.: Sexual differences in adult skulls of *Pan troglodytes. J Human Evol, 3*:123, 1974.

Gans, B. J. and Sarnat, B. G.: Satural facial growth of the *Macaca rhesus* monkey: a gross and serial roentgenographic study by means of metallic implants. *Am J Orthod, 37*:827, 1951.

Gavan, J. A.: Blood group factors in anthropoid apes and monkeys. *Am J Phys Anthropol, 11*:39, 1953.

Gilda, J. E.: Analysis of linear facial growth. *Angle Orthod, 44*:1, 1974.

Giles, E.: Cranial allometry in the great apes. *Hum Biol, 28*:43, 1956.

Glanville, E. V.: Nasal shape, prognathism and adaptation in man. *Am J Phys Anthropol, 30*:29, 1969.

Groves, C. P. and Humphrey, N. K.: Asymmetry in gorilla skulls: evidence of lateralized brain function? *Nature* (Lond), *244*:53, 1973.

Harris, J. E. and Kowalski, C. J.: Relationship between tooth size, body size and craniofacial dimensions in young adults. *Am J Phys Anthropol, 41*: 484, 1974.

Harris, J. E., Kowalski, C. J., and Watnick, S. S.: Genetic factors in the shape of the craniofacial complex. *Angle Orthod, 43*:107, 1973.

Hassund, A. and Sivertsen, R.: Dental arch space and facial type. *Angle Orthod*, 41:140, 1971.

Heflin, B. M.: A three-dimensional cephalometric study of the influence of the midpalatal suture on the bones of the face. *Am J Orthod*, 57:194, 1970.

Hemmer, H.: Allometrische untersuchungen am Schadal von *Homo sapiens* unter besonderer Beruchsichtung des Brachykephalisations-Problems. *Homo*, 17:190, 1967.

Hertzog, K. P.: Association between discontinuous cranial traits. *Am J Phys Anthropol*, 29:397, 1968.

Hooton, E. A.: Up from the Ape. New York, Macmillan, 1946.

Howell, F. C.: European and northwest African Middle Pleistocene hominids. *Curr Anthropol*, 1:1, 1960.

Howells, W. W.: Analysis of patterns of variation in crania of recent man. In Tuttle, R. (Ed.): *The Functional and Evolutionary Biology of Primates*. Chicago, Aldine, 1972.

Hrdlicka, A.: New data on the teeth of early man and certain European fossil apes. *Am J Phys Anthropol*, 7:109, 1924.

Hunt, E. E.: The continuing evolution of modern man. *Cold Spring Harbour Symp Quant Biol*, 24:245, 1960.

Hunter, W. S. and Garn, S. M.: Disproportionate sexual dimorphism in the human face. *Am J Phys Anthropol*, 36:133, 1972.

Huxley, J. S.: *Problems of Relative Growth*. London, Methuen, 1932.

Ingervall, B. and Lennartsson, B.: Facial skeletal morphology and dental arch dimensions in girls with postnormal occlusion. *Odont Revy*, 23:63, 1972.

Ingervall, B. and Lewin, T.: Stability of relations between cranial vault and cranial base during secular changes in external head dimensions. *Acta Morphol Neerl Scand*, 12:215, 1974.

Johnson, C. C., Gorlin, R. J., and Anderson, V. E.: Torus mandibularis. *Am J Hum Genet*, 17:433, 1965.

Keith, A.: Cranial characters of gorillas and chimpanzees. *Nature* (Lond), 120:914, 1927.

Kemble, J. V. H.: The importance of the nasal septum in facial development. *J Laryngol Otol*, 87:379, 1973.

Kitson, E.: A study of the negro skull with special reference to the crania from Kenya colony. *Biometrika*, 23:271, 1931.

Knip, A. S.: Metrical and non-metrical measurements on the skeletal remains of Christian populations from two sites in Sudanese Nubia I and II. *Proc Kon Ned Akad v Wetensch Amsterdam*, 73:433, 1970.

Kraus, B., Wise, W. and Frie, R.: Heredity and the cranifacial complex. *Am J Orthod*, 45:172, 1959.

Krogman, W. M.: The problems of growth changes in the face and skull of anthropoids and man. *Dent Cosmos*, 72:624, 1930.

Krogman, W. M.: Studies in growth changes in the skull and face of anthropoids. *Am J Anat*, 47:89, 325, 343, 1931.

Krogstad, O.: The relationship between the lower margin of the nasal aperture and the maxillary alveolar process. *Z morphol Anthropol*, 65:34, 1973.

Landauer, C. A.: A factor analysis of the facial skeleton. *Hum Biol*, 34:239, 1962.

Laughlin, W. S. and Jorgensen, J. B.: Isolate variation in Greenlandic Eskimo crania. *Acta Genet Statist Med*, 6:3, 1956.

Lindegard, B.: Variations in human body build. *Acta Psychiatr Suppl*, 86:1, 1953.

Manera, J. F. and Subtelny, J. D.: A cephalometric study of growth of the nose. *Am J Orthod*, 47:703, 1961.

Martin, R.: *Lehrbuch der Anthropologie*. Jena, Gustav Fischer, 1928.

McGown, T. D. and Keith, A.: *The Stone Age of Mount Carmel*. Oxford, Clarendon Press, 1939.

Moore, W. J.: Head growth of the Macaque monkey. *Am J Orthod*, 35:654, 1949.

Moore, W. J. and Lavelle, C. L. B.: *Growth of the Facial Skeleton in the Hominoidea*. London, Acad Press, 1974.

Moore, W. J., Adams, L. M., and Lavelle, C. L. B: Head posture in the Hominoidea. *J Zool* (Lond), 169:409, 1973.

Morant, G. M.: A study of Egyptian craniology from prehistoric to Roman times. *Biometrika*, 17:1, 1925.

Morant, G. M.: A first study of the craniology of England and Scotland from Neolithic to early historic times. *Biometrika*, 18:56, 1926.

Morant, G. M.: Studies of Paleolithic man. *Ann Eugen* (Lond), 3:337, 1928.

———: Studies of Paleolithic man. *Ann Eugen* (Lond), 4:109, 1930.

Moss, M. L. and Young, R. W.: A functional approach to craniology. *Am J Phys Anthropol*, 18:281, 1960.

Park, A. W. and Nowosielski-Slepowron, B. A. J.: The effects of litter size on rat skull growth. *Acta Anat*, 81:541, 1972.

Pearson, K. and Woo, T. L.: Further investigation of the morphogenetic characters of the individual bones of the human skull. *Biometrika*, 27:424, 1935.

Polyak, S.: *The Vertebrate Visual System*. Chicago, U Chicago U Pr, 1958.

Posen, J. M.: A longitudinal study of the growth of the nose. *Am J Orthod*, 53:746, 1967.

Preston, C. B. and Evans, W. G.: The cephalometric analysis of Cercopithecus aethiops. *Am J Phys Anthropol*, 44:105, 1976.

Riesenfeld, A.: The variability of the temporal lines, its causes and effects. *Am J Phys Anthropol*, 13:599, 1955.

Riesenfeld, A. and Siegal, M. I.: The relationship between facial proportions and root length in the dentition of dogs. *Am J Phys Anthropol,* 33:429, 1970.

Rightmire, G. P.: Cranial measurements and discrete traits in distance studies of African negro skulls. *Hum Biol, 44*:263, 1972.

Robinson, J. T.: The dentition of the Austral opithecinae. *Transv Mus Mem* No. 9, 1956.

Sauter, M. R.: A propos de l'architecture de l'occipital. *Anthropol Paris,* 50: 469, 1941-46.

Schultz, A. H.: The skeleton of the trunk and limbs of higher primates. *Hum Biol, 2*:203, 1930.

Schultz, A. H.: Growth and development of the chimpanzee. *Contrib Embryol Carneg Inst, 28*:1, 1940.

Schultz, A. H.: Growth and development of the orangutan. *Contrib Embryol Carneg Inst, 29*:59, 1941.

Schultz, A. H.: Age changes and variability in gibbons. *Am J Phys Anthropo!, 2*:1, 1944.

Schultz, A. H.: The physical distinctions of man. *Proc Am Phil Soc,* 94: 428, 1950a.

Schultz, A. H.: Morphological observations on gorillas. In *The Henry Cushier Raven Memorial Volume.* New York, Columbia U Pr, 1950b, pp. 227-251.

Schultz, A. H.: The position of the occipital condyles and of the face relative to the skull base in primates. *Am J Phys Anthropol, 13*:97, 1955.

Schultz, A. H.: Postembryonic age changes. *Primatologia* (Basel), *1*:887, 1956.

Senyurek, M. S.: Cranial equilibrium index. *Am J Phys Anthropol, 24*:23, 1938.

Solow, B.: The pattern of craniofacial associations. *Acta Odont Scand,* 24: suppl 46, 1966.

Steegman, A. T.: Cold adaptation and the human face. *Am J Phys Anthropol, 32*:243, 1970.

Stockard, C. R.: Human types and growth reactions. *Am J Anat, 31*:261, 1923.

Tallgren, S.: Changes in adult face height due to ageing, wear and loss of teeth and prosthetic treatment. *Acta Odont Scand, 15*:1, 1957.

Thurow, R. C.: The decimal calendar. *Angle Orthod, 31*:69, 1961.

Tobias, P. V.: Cranial capacity of *Zinjanthropus* and other australopithecines. *Nature* (Lond), *197*:743, 1963.

Tobias, P. V.: *Olduvai Gorge.* London, Cambridge U Pr, 1967, Vol. 2.

Todd, T. W.: Prognathism; a study in development of the face. *J Am Dent Assoc, 19*:2172, 1932.

Trevor, J. C.: The physical characters of the Sandawe. *J R Anthropol Inst,* 77:61, 1947.

Van Melsen, A. G. .: The morphogenesis of the human head. *Acta Morph Neerl Scand,* 10:3, 1972.
Vig, P. and Hewitt, A. B.: Is craniofacial asymmetry and adaptation for masticatory function an evolutionary process? *Nature* (Lond), 248:165, 1974.
Weidenreich, F.: Uber die pneumatischen Nebenraume des Kopfes. *Z Anat Entw Gesch,* 72:55, 1924.
Weidenreich, F.: The skull of *Sinanthropus pekinensis. Palaeont Sin,* 127:1, 1943.
Weidenreich, F.: Giant early man from Java and South China. *Anthropol Pap Am Mus Nat Hist,* 40:1, 1945.
Weidenreich, F.: *Apes, Giants and Man.* Illinois, Chicago U Pr, 1946.
Weiner, J. S.: Nose shape and climate. *Am J Phys Anthropol,* 12:615, 1954.
Weinert, H.: Die Ausbildung der Stirnhohlen als stammesgeschtliches Merkmal. *Z Morph Anthropol,* 25:365, 1926.
Weinert, H.: Der Urmenschenschadel von Steinheim. *Z Morph Anthropol,* 35:463, 1936.
Wolpoff, M. R.: Climatic influence on the skeletal nasal aperture. *Am J Phys Anthropol,* 29:405, 1968.
Wolpoff, M. H.: Sagittal cresting in the South African anstralopithecines. *Amer J Phys Anthropol,* 40:397, 1974.
Woo, T. L.: On the asymmetry of the human skull. *Biometrika,* 22:324, 1931.
Woo, T. L. and Morant, G. M.: A biometric study of the "flatness" of the facial skeleton in man. *Biometrika,* 26:196, 1934.
Woollard, H. H.: The retina of primates. *Proc Zool Soc,* 1:1, 1927.
Wylie, W. L.: Mandibular incisor—its role in facial esthetics. *Angle Orthod,* 25:32, 1955.
Yamazaki, K.: Quelques probemes du visage humain. *Rev Stomat* (Paris), 69:695, 1969.
Young, R. W.: The influence of cranial contents on postnatal growth of the skull in the rat. *Am J Anat,* 105:383, 1960.
Zuckerman, S.: Growth changes in the skull of the baboon, *Papio porcarius. Proc Zool Soc* (Lond), 2:843, 1926.

CHAPTER THREE

THE PRIMATE MASTICATORY SYSTEM

INTRODUCTION

THE BASIC PRINCIPLES underlying the contrast between different primate jaws are considered in this chapter. This is primarily due to the fact that there is a dearth of quantitative data relating to the primate jaws, presumably arising from the lack of homologous datum points, and as a consequence, primate taxonomy based on jaw form will remain largely subjective for the present. Indeed, from a familial study, Harris and Kowalski (1976) concluded that it is as yet impossible to ascertain the exact proportions of genetic and environmental factors in the control of craniofacial growth in man. Furthermore, until this has been resolved for other primates as well, the interpretation of evolutionary changes will remain impeded.

The size and power of the masticatory apparatus affects head form (Ringqvist, 1973), in that a large maxilla requires a large mandible resulting in a prognathous animal if the skull is not equally expanded. Also, since all the muscles of mastication have origin on the skull, the extent of their attachment influences skull form (Kurisu, Kato, and Kauda, 1972). Masticatory stress is contained within the perimeter of the skull, so that masticatory power will affect the thickness and form of the facial bones (Boyd, Castelli, and Huelke, 1967). Due to the location of the nasal cavity and orbits, however, masticatory stress must be carried around these cavities by the maxilla and malar bones, with the supraorbital torus acting as a cross-member at the top of the skull.

Many comments have been made on the mechanical significance of the facial skeleton of the gorilla and various characteristics and differences compared with the human facial skeleton, e.g. the split-line analysis by Tappen (1954). Enlow (1973) notes that the strain distribution in the gorilla facial skeleton—as determined by wire strain gauges—is fundamentally identical to that of the human facial skeleton. The supraorbital and infraorbital regions can be regarded as a kind of beam in a rigid-frame structure which is acted upon by the bending moment due to muscular action. In the human skull, both regions are thick and accept strong bending moments. In the gorilla skull, the supraorbital region is thin compared with the infraorbital region. The latter, therefore, theoretically accepts the majority of the bending movement. This may be the reason for the fact that the thinner supraorbital region is not intensively strained as compared with the infraorbital region (Enlow, 1973). Moreover, the differences in the distribution of the strain in the facial skeleton seem to show resemblances with the differences in the split-line patterns reported by Tappen (1954). It is obvious, however, that further mechanical studies are required in the primate skull before the differences in structure can be functionally analyzed.

Little compressive forces develop between the teeth of reptiles, whose jaws function mainly as food traps. Little more than a grasping action occurs with the jaws of most primitive primates (tree shrews), although as the grinding action of the teeth evolved more power developed with the resultant marked change in mandibular shape. Subsequent to the evolution of a more efficient and powerful masticatory action, the whole of the craniofacial skeleton was adapted to accommodate its functional needs. This illustrates the fact that evolutionary changes cannot be considered in one isolated region of the skull, but ramifications in other regions must also be taken into account. This is demonstrated by the work of Bilsborough (1972) who showed that some hominid cranial complexes (upper face, mandible, balance and cranial vault) have changed considerably during the Pleistocene, whereas others (articular region and upper jaw) exhibited little change over the same period.

In addition to the masticatory muscles, prognathism affects facial form. The degree of prognathism may be quantified (Laffout and Aaron, 1961), and reveals marked contrasts between man and apes. For instance, the facial angle, defined as the angle between the Frankfort Horizontal and the line joining nasion to prosthion, ranges between 41 to 56 degrees for Pongidae (Tobias, 1967), and 77 to 89 degrees for *Homo sapiens*. Also, the nasal profile angle, defined as the angle between the Frankfort Horizontal and the lining joining nasion to nasospinale, has a range of 81 to 90 degrees for *Homo sapiens* and 51 to 60 degrees for Pongidae (Tobias, 1967). Thus, the degree of prognathism which possibly reflects the degree of development of the masticatory apparatus has a bearing on overall facial form.

During evolution of the primate masticatory apparatus, two important changes have occurred. The mandibular condyle is more or less on the same plane as the occlusal surface of the teeth in primitive primates, whereas in higher primates, it has moved well above this plane. As a result of this shift, in place of the situation in lower primates where the molars occlude first followed by the incisors, the whole of the dentition occludes simultaneously in higher forms. The increased distance between condyle and occlusal plane permits lateral movement necessary for grinding, so that in these higher forms the mandibular ramus tends to ascend rather than lie in a horizontal plane. In the skull itself, displacement of the teeth relative to the mandibular articulation has been accommodated by a lowering of the nasal cavity floor and dental arcade relative to the neurocranium and orbits, and in some species the palate is vaulted.

The power developed between the molars depends upon the length of the power arm and load arm of the mandible for a given volume of muscle. Thus, the shorter the load arm, i.e. the distance between the teeth and mandibular condyle projected onto the occlusal plane, the greater the compressive power developed between the teeth. During primate evolution, the relative shortening of the load arm has been achieved mainly by the posteroinferior location of the dental arcades relative to the neurocranium.

Thus, during primate evolution, marked changes have occurred in the whole of the masticatory apparatus, although there is a dearth of quantitative data relating to such changes.

The phylogenetic development of the mammalian masticatory system is summarized in most standard zoological texts, to which the reader is referred for further information. The development of a dentary-squamosal (temporal) articulation, which, in primitive mammals, is in conjunction with the more primitive articular-quadrant articulation, indicates that the former joint does not serve as either a hinge or fulcrum. The function of the dentary-squamosal articulation probably relates to the need for a mechanism to locate the mandible. Such a device would be necessary in view of the complexity of mandibular function following the development of a differentiated tooth row. The loss of the primitive reptilian articular-quadrate articulation that was suitable as a load-bearing fulcrum indicates that, in mammals, the jaw joint is not load-bearing. Possibly the only forces acting through the temporomandibular joint are those required to maintain the integrity and alignment of the osseous components. The eradication of a load-bearing joint would necessitate the differentiation and refinement of the masticatory musculature so that the forces produced by it were in mechanical equilibrium with the bite force generated at the tooth row. This implies that the forces involved can be demonstrated to form a closed system of vectors and such a differentiation of the musculature is seen in the development of the complex system of mammalian jaw adductors.

Alternatively, the bite forces, irrespective of their position on the tooth row, may be directed toward the apex of a mechanical framework formed by the bony architecture of the facial skeleton. This model assumes differential contraction of various components of the masticatory musculature, depending on the position of the bite point on the tooth row. The facial framework is generally a subpyramidal structure, the base being represented by the alveolar process and to some extent by the pterygoid laminae and the anterior part of the cranial base. The bony roots of the canines serve as structural members, being surrounded by a strut of bone which is less cancellous than other parts of the maxilla which are not concerned with the general framework. The two structural members represented by the canine roots meet at the nasion and are continued as a single strut between the orbits to become confluent with the supraorbital tori and glabella. Inferior to the lateral margin of the orbit, at thickening of the alveolar and maxillary bone signifies the base of another intermediate structural member of the facial framework. This member continues upward as the posterior orbital bar and the supraorbital torus to become confluent with the anterior struts previously described. The temporal and maxillary components of the zygoma comprise a horizontal structural brace which provides considerable mechanical strength to the intermediate vertical struct. A more complex arrangement forms the posterior struts of the framework. It incorporates the wings of the sphenoid bone and the anterior cranial base. Similarly, this also becomes confluent with the apex of the facial pyramid in the region superior to the supraorbital tori. This framework is assumed to be optimal for the absorption of forces generated at its base when those forces are directed toward the apex of the pyramid. Forces not so directed are split up into components, one of which is directed towards the apex.

Throughout man's phylogenetic evolution, there has been a tendency to specialize the facial framework and the masticatory musculature such that an efficient posterior bite is attained. The specialization has progressed to the point of incorporating most of the tooth row into a posterior type of bite, although the introduction of overjets in recent populations may upset this functional trend. In *Australopithecus*, the facial framework is nearly vertical, whereas in *Gorilla, Pan,* and *Macaca*, the framework is oriented obliquely backward with the apex above and behind the base. In *Homo*, the framework tends to be inclined obliquely forward with the apex above and in front of the base. The facial framework of *Australopithecus* is therefore *tucked* further beneath the cranium than in monkeys and apes, but less than in modern man. The degree to which the facial skeleton is tucked beneath the cranium (klinorhynchy) forms a phylogenetic trend in hominids. Such a trend is related to the major development and expansion of the frontol lobes of the cerebral hemispheres and to the increasing use of the dentition for grinding, together with the increasing development of manual dexterity. Inclination of the facial skeleton in this manner leads to a change in the relationship between the cranium and facial skeleton and masticatory musculature, such that the vertical components of the masticatory forces are enhanced and a powerful posterior bite results. Furthermore, the forward rotation of the apex of the facial framework has resulted in the reorientation of the frontal bone. In lower primates, the parasagittal axis of the supraorbital torus tends to form an acute angle with the base of the cranium, whereas in man the same angle is obtuse. The *browridge* is therefore inclined forward.

There is a phylogenetic trend for the craniofacial axis to become increasingly flexed and to prevent undue encroachment of the pharyngeal contents, the maxillary part of the facial skeleton has tended to become relatively smaller. The maxillary dentition does not seem to have followed the trend for reduction at the same rate. This is possibly one of the reasons for dental overcrowding and could conceivably be a causative factor in the development of an arched tooth row.

The mandible appears to develop independently, i.e. it develops in a distinct soft-tissue matrix compared with the maxilla. Nevertheless, in order for normal masticatory function, the development of the maxilla and mandible must, to a certain extent, be interrelated one with another.

Although the maxilla may be hyperplastic when excessive klinorhynchy occurs, the sense organs contained within the face appear to develop normally. When excessive klinorhynchy develops, therefore, the orbits become crowded into the reduced facial area. The result is that there is a certain amount of distortion of the facial framework leading to an adverse distribution of the bite force vectors. This emphasizes, therefore, that the jaws are intimately related with the facial skeleton as a whole, the latter being again interrelated with the neurocranium. The skull is, therefore, a complex biological entity rather than a series of discrete biological units.

MAXILLA

With the exception of the premaxilla (Anderson, 1903, 1908; Montagu, 1935), the primate jaw has received scant attention by physical anthropologists.

The most outstanding work in this respect is by Bilsborough (1971) who measured the upper jaws of fossil and extant hominids and pongids. The major contrasts between modern taxa were in the anterior region of the maxilla, where considerable differences existed between the pongid genera, while *Homo sapiens* was markedly different from them all. In general, the variability in pongids was greater than in *Homo sapiens*, due to marked sexual dimorphism in the former. In character means, *Dryopithecus major* was intermediate between *Pan* and *Gorilla*, whereas *Ramapithecus* resembled *Pan* rather than any other extant hominoid. *Dryopithecus africanus* was closer to *Homo sapiens* than to modern apes, although the anterior dentition and shape of the dental arcade was pongid rather than hominid.

Canonical analysis showed considerable discrimination between the groups. *Homo sapiens* was the modern genus closest to *Dryopithecus africanus*, with *Pan* only slightly further away. In *Pan*, this reflected a basic similarity in the overall proportions of an absolutely larger maxillary region, whereas the human sample, although contrasting in palatal shape with *Dryopithecus africanus*, approached it more closely in size than any extant pongid.

The separation between *Dryopithecus africanus* and *Gorilla* was mainly the result of size differences, whereas that between *Dryopithecus africanus* and *Pongo* was mainly due to size together with shape differences resulting from alveolar prognathism and incisor hypertrophy in the modern genus.

The distinctive features of the human maxilla are reflected by the separation of *Homo sapiens* from the extant pongids. The maximum divergence occurs between *Homo sapiens* and *Pongo*, which reflects the morphological contrasts between these two, especially in the degree of alveolar prognathism, size of the anterior teeth and shape of the dental arcade. The African pongids are closer to one another than either is to *Pongo*, with the separation between *Pan* and *Gorilla*, reflecting size differences, and that between *Pan* and *Pongo*, and between *Pongo* and *Gorilla* being mainly due to contrasts in jaw shape. *Dryopithecus major* is closer to *Gorilla* than to any extant hominid, whereas the great separation between *Dryopithecus africanus* and *Dryopithecus major* are mainly due to size. *Ramapithecus* is nearer to *Pan*, although *Dryopithecus africanus* and *Homo sapiens* are also relatively close resulting in the proportions of the alveolar incisor region.

The typical pongid dental arcade is present in Miocene

dryopithecines, although there are numerous differences between fossil and modern genera, especially in the anterior palatal region. All African dryopithecines have relatively narrow incisors compared with extant genera. Of the extant pongids, *Gorilla* has the smallest incisors and largest cheek teeth relative to palatal size, whereas *Pongo* has the largest incisor region. This is reflected by the fact that *Pongo* is more frugivorous while *Gorilla* is predominantly herbivorous.

The functional basis of the hominid parabolic dental arcade and the selective factors underlying it are obscure. A restriction of the anterior dentition certainly occurred, but other factors, such as a reduction in anterior cranial length due to an adaptation to orthograde posture are probably involved. On the basis of these findings, however, Bilsborough (1971) concluded that *Dryopithecus africanus* was the most likely common ancestor of hominids and pongids, with *Dryopithecus major* ancestral to *Gorilla*. This research now needs to be extended to include other primate groups.

The maxilla, or more appropriately, the nasomaxillary complex, comprises a series of bones or a group of areas including the maxillary arch, palatine process, premaxillary area, zygomatic process, nasal region, maxillary sinuses, and orbital floor. These membranous bones have sutural processes which are present at the junction between the maxillary complex and other cranifacial bones to which the maxillary complex is attached.

Maxillary growth is complex (Enlow and Bang, 1965), although in man there is a general downward and forward movement as the maxilla is presumably relocated downward and forward as the posterior regions impinge on the cranial base. The maxillary arch also elongates, moves in a posterior direction and increases in height to accomodate the permanent dentition. As the maxillary arch increases in height, the palatal floor compensatorily moves inferiorly so that the nasal chamber is enlarged vertically as the palatal floor moves inferiorly. In addition, the premaxillary region responds to palatal growth by shifting downward. The zygomatic process moves posterolaterally as the maxilla elongates, and in addition, the sutures in the zygomatic region are associated with anterior displacement of the entire maxilla. The frontal processes of the zygomatic bone move in an anterior superior and lateral direction, and the orbital floor is also moved in the same direction, so that the zygomatic bone effectively enlarges in all directions. Growth of the nasal bones and maxillary aspects of the lateronasal walls proceed anteriorly and the maxillary sinus apparently exists to fill the space previously occupied by the facial bones.

With the exception of the work of Enlow (1968), there is a dearth of data relating to growth of the maxilla of primates other than man.

The growth patterns found in the anterior portion of the maxillary arch in man and *Macaca* (Enlow, 1968) are basically dissimilar. The arch of the maxilla anterior to the malar region is characteristically depository in the monkey, which contrasts with the typically resorptive nature of the anterior part of the human maxilla. The external surface is convex in the monkey, but uniquely concave in the human maxilla. In both species, a predominantly downward direction of growth occurs in this region, although in the monkey, the forward portion of the maxillary arch, including its separate premaxillary bones, grows in a distinct forward course as well. The contrasting pattern of growth produces an essentially downward pattern of maxillary arch growth in man, but a downwards and forwards pattern in the monkey. This growth combination in the monkey produces a premaxillary region that protrudes noticeably forward of the nose. In the postnatal human face, loss of the premaxillary sutures, lack of forward premaxillary cortical growth, and diminutive anterior displacement associated with sutural growth at the various zygomatic and maxillary sutures result in a marked decrease in maxillary prognathism. The anterior mode of bony and cartilaginous growth in the overlying nasal region brings about the formation of the characteristic human nose which extends well forward of the short maxillary arch.

In both monkey and man, the marked downward rather than primarily forward growth of the entire nasal area is also a critical factor associated with decreased prognathism. Reduced dentition in the monkey further contributes to a facial profile approaching that of man. In the human skull, dental reduction and the distinctively resorptive nature of the anterior maxillary and mandibular arches together with the essentially vertical orientation of the face as a whole, are primarily responsible for the extreme lack of facial prognathism. Nevertheless, lack of quantitative growth data relating to other primates inhibits the exploitation of the maxilla in taxonomy.

MANDIBLE

With evolution of the forelimb for prehensile functions, there was a concommittant reduction in the size of the jaws and muscles of mastication. Diminution of jaw size was also associated with snout reduction concommittent with the diminishing importance of the nasal cavities (Crompton, 1963).

In tree shrews, the mandible comprizes a horizontal body and a vertical ramus terminating with the coronoid process and articular process whereby the mandible articulates with the skull. The lower border of the mandible is extended backwards into a hook-shaped angular process, whereas the two sides of the mandible are joined together by the symphysis menti. The mandible of modern lemurs commonly shows a slender body, with the symphysial region slanting strongly forwards in conformity with the procumency of the incisors and lower canines. The vertical ramus is relatively low, with well-marked coronoid and angular processes. The two halves of the mandible diverge from the symphysis only to a slight degree. The symphysis menti is normally unossified in the adult, although fused in the fossil lemurs, *Adapis* and *Archaeolemur*.

In tarsoids, the mandible is lightly built, with relatively low and broad ramus and a small coronoid process. The bodies diverge conspicuously from the region of the symphysis, and this is correlated with the relatively wide cranial base which separates the glenoid fossae. In modern Tarsius, the symphysis menti remains unossified, but in the mandible of Middle Eocene *Caenopithecus*, it is synostosed, thereby resembling the normal condition in the Anthropoidea. The mandible is stoutly built in most monkeys, apes, and man, and with recession of the snout, the corpus tends to be shorter relative to the vertical ramus, with the two halves firmly synostosed at the symphysis menti.

Although numerous metrical descriptions have been provided for the human mandible (Harrower, 1928; Martin, 1936; Morant, 1936; Cleaver, 1937; Hrdlicka, 1940) descriptions of the mandibles of other primate groups have remained qualitative. Thus, apart from gross trends, it is difficult to describe the evolutionary trends of the primate mandible objectively. The main drawback to metrical studies of the primate mandible is the lack of homologous datum points. For instance, Murphy (1957) has remarked that since the pogonion is lacking in all but the human mandible, no comparison can be effected between the mental angle of human and nonhuman primates. Nevertheless, until metrical descriptions of the primate mandible have been provided, analysis of evolutionary changes will remains objective, as will primate taxonomy when based solely on this bone.

Growth of the human mandible has been studied by many workers (e.g. Hunter, 1837; Humphrey, 1863; Brodie, 1942; Baume, 1953; Bjork, 1963; Enlow, 1968; Enlow and Harris, 1964), but whether all primates share a common mandibular growth pattern has yet to be elucidated. During growth, the human mandible enlarges and relocates in space so that it moves in many directions at the same time. The growth process involves synchronous and selective deposition and resorption of bone from membrane surfaces coupled with interstitial and appositional growth changes in the condyle. The main growth thrust appears to be in an upward and backward direction involving bony deposition along the posterior aspects of the ramus and in the condylar area. The condylar neck is continually relocated in a superior-posterior direction as it follows the moving condyle. The more basal parts of the ramus move labially in common with the contralateral lingual side. The coronoid process moves in a lingual, superior-posterior direction at the same time, and the anterior part of the ascending ramus becomes flattened and is converted into the lengthening corpus. In addition, the lengthening mandibular body is relocated mesially and as a result of this shift, the molars, which erupt in the tuberous area, are aligned with the rest of the dentition. This remodelling pattern then changes in the area of the premolars and the entire lingual surface of the corpus moves toward the midline. The labial surface moves buccally except in the alveolar region between the canines. This reversal creates a flattened area in the alveolar zone between the canines. How far this growth pattern is mimicked by other primate groups awaits the accumulation of other quantitative data.

Enlow (1968) indicated that many of the areas of the mandible characterized by periosteal resorption and endosteal deposition also have muscle attachments. This queries the concept that the gross morphology of a bone is a direct consequence of periosteal growth stimulated by muscle tension (Avis, 1961; Barber, Green and Cox, 1963). Moss and Salentijn (1969) similarly stated that variation in bony resorption and deposition in relation to functional stresses are not necessarily related to muscle attachments. Moss (1960) and Moss and Rankow (1968) contend that mandibular growth is a combination of the effects of both the capsular and periosteal matrices. The expansion of the orofacial capsule occurs as the capsular matrix grows in response to the functional demands, e.g. patency of the airway and digestive tract. The enclosed mandible is passively and secondarily translated downward and forward in space as the capsular matrix increases in size. Movement of the mandible away from the cranial base, therefore, would tend to cause a separation of the elements of the temporomandibular joint, so that the condyle grows of necessity to maintain the joint in functional contact, i.e. condylar growth is secondary and compensatory. The periosteal matrices related to the constituent microelements of the mandible also respond to the volumetric expansion. Inevitably, they grow as their spatial position is altered and their functional demands change accordingly. The altered functional requirements elicit changes in size and/or shape of the microskeletal units. Hence, movements of the mandible in space result from volumetric expansion of the demands of the periosteal matrices. This comprises the main directive process involved in mandibular growth. Moreover, Moss's functional matrix theory contends that the functioning tissues, including muscles, nerves, blood vessels, glands and teeth, coupled with the oral, nasal and pharyngeal cavities, comprise the primary forces controlling the spatial position and size of the mandible.

It is obvious that the various functional matrices will vary in different primates and this may account for variation in mandibular form. Nevertheless, while Moss's functional matrix theory remains attractive, further data are required before such a notion can be applied to all primate groups.

There is evidence that during recent times in Europe, there has been a reduction in the dimensions of the mandible without a corresponding diminution in tooth dimensions (Lysell, 1958). Although the reasons for these changes are obscure, it is traditional to assume that the reduction in mandibular dimensions results from a diminution in the physical consistency of the diet, i.e. mastication of soft diets provides little stimulus for growth of the facial skeleton (Watt and Williams, 1951). Another theory contends that the reduction in mandibular dimensions is part of a long-term evolutionary trend leading to a reduction in the entire facial skeleton and is independent of any change in dietary consistency (Hooton, 1946). If dietary changes were primarily responsible for the reduction in mandibular dimensions, then the mandibular ramus (which provides for the insertion of the muscles of mastication) would be more reduced in size compared with the corpus of the mandible (which bears the teeth).

In order to investigate this phenomenon, Lavelle (1972) compared the Romano-British, Anglo-Saxon and nineteenth century British mandibles. Using multivariate analysis, no significant difference in the degree of discrimination was obtained from analysis of either the dimensions of the ramus or corpus of the mandible. This is not surprising when Moss's functional matrix theory is considered. The mandible is not a unitary biological entity, but a composite of several relatively independent functional components. The skeletal units corresponding to these functional mandibular components include the alveolar process, coronoid process, angular process, corpus, condylar process, and chin. Moreover, it is experimentally demonstrable that the functional matrix is primary and that the presence, size, shape, spatial position and growth of any skeletal unit is secondary, compensatory and mechanically obligatory to changes in the size, shape and spatial position of its related functional matrix. In the mandible, the teeth form the functional matrix of the alveolar skeletal unit and the temporalis muscle is the matrix related to the coronoid process. Thus, the mandible appears to comprise a number of individual functional cranial components each with its functional matrix and skeletal unit. Individual functional cranial components may have no causal relationship to each other, so that any alteration of one matrix causes a corresponding alteration of its specific skeletal unit alone, and without necessarily causing an alteration in other functional cranial components. It is reasonable to suppose, therefore, that any evolutionary change would be complex rather than merely being reflected in the mandibular corpus or ramus dimensions. Multivariate statistical analysis on many mandibular dimensions is imperative in order to elucidate evolutionary changes in the primate mandible. But, as previously stated, other than for man, there is a dearth of quantitative data relating to the primate mandible.

The form and bulk of the masseter muscle is related to the extent of its area of attachment, i.e. to the zygomatic arch and ramus of the mandible, (Scott 1957). Furthermore, Scott has demonstrated that the breadth of the mandibular ramus is related to the degree of development of the muscles of mastication, with the latter in turn being related to molar size. This worker devised an index M/R, where M is the combined mesiodistal length of the lower permanent molars and R is the breadth of the mandibular ramus. Scott (1957) computed the range of this index for hominids to be 70 to 107 percent, with that of *Pan* and *Gorilla* being 61 to 62 percent. The index for the Heidelberg mandible, with its considerable breadth of mandibular ramus has a value of 70 percent. As with most indices, however, there is as yet insufficient data relating to other primate groups for undue reliance to be placed on this index for taxonomy.

Angle of the Mandible

There is a pronounced angular process of the mandible of prosimians, from the earliest Paleocene representatives to those of the present day; whereas no such structure is present in Old World moneys or apes. Nevertheless, a morphological feature which has been present in primates for at least 65 million years must have functional significance and adaptive value. This supposition is strengthened by the fact that an angular process was a phenonemon common to early Tertiary mammals and also characterized even earlier, supposedly ancestral, mammalian forms as far back as the Middle Mesozoic. In Old World monkeys and apes, the angle of the jaw is greatly reduced and this altered structure must represent a different kind of adaptive mechanism, which also lasted for some 30 million years.

Since both fossil and modern prosimians share with the rat well-developed anterior teeth and shallow jaws with an angular process, this bone-muscle-tooth complex should share a similar function. Among living forms, even the aye-aye has rodent incisors. All remaining modern prosimians, except *Tarsius*, have procumbent incisors. These are part of an important adaptive complex in shallow-jawed prosimians for grooming. In insectivores and tree shrews, there seems to be an adaptation in which the lower incisors are used against the back of the upper incisors. All evidence suggests that in many mammals, procumbent lower incisors were evolved as a cropping and cutting adaptation with the forward and upward motion of the jaw in these forms accomplished by the superficial masseter and internal pterygoid muscles being attached to an enlarged angular region of shallow jaws.

Hence, the significance of the angle of the mandible appears to reflect the functional bone-muscle complex. For example, during evolution, as the jaw became deeper, the superficial masseter-internal pterygoid muscles became more of an elevator and less of a protruder of the jaw. As the jaws became deeper, the line of action of these two muscles became more vertical. This transition was probably correlated with the emergence of larger molars leading to a more efficient grinding apparatus. In man, monkeys, and apes, the deepening of the mandible was probably associated with the emergence of the external pterygoid muscle acting mainly as a mandibular protruder. Thus, the morphology of the angle is associated with the action of the muscles of mastication, although in view of the complex muscular insertions in this area, the interpretation of variation in the angle of the mandible in hominoids requires further data.

The Chin

Although absent in most primates, innumerable descriptions of the "chin" have been provided for fossil and recent man (e.g.

Robinson, 1914; Weidenreich, 1936; Hershkovitz, 1970). There is also an extension on the posterior aspect of the mental symphysis —the simian shelf—which is so well developed in monkeys, anthropoids and dryopithecine fossils, although absent in man and dubious in australopithecines (Broom and Robinson, 1952). Although the simian shelf has some obvious mechanical function relating to jaw strength, the significance of the chin remains controversial. Numerous theories relating to the evolution of the chin have been reviewed by DuBrul and Sicher (1954). These workers reject the "shifting" theory of Bolk (1924) where speech was assumed the stimulus for chin development, Virchow's (1910) claim that the chin resulted from the function of the facial muscles and Wegener's (1927) view that the chin was initiated by the action of the temporalis muscle. They also rejected the notion of Weidenreich (1934) that the chin developed in a passive manner when tooth size decreased during evolution and the alveolar portion of the mandible was reduced, i.e. as the primate mandible changed from a pointed to a broad and flattened arch accompanied by a more erect dentition, the resultant buckling of the symphysis left a mechanically formed triangular deficiency owing to the increasing separation of the basal parts, and this triangular gap was filled by the mental trigonium. Instead, DuBrul and Sicher (1954) enlist the action of the external pterygoid muscles as the stimulus for chin formation. These latter workers noted that in long-headed lower primates, the external pterygoid muscles form only a small angle with the saggital plane, while the angle is much greater in short-headed man. Consequently, these muscles are more nearly transversely oriented so that they squeeze the two condyles together producing stresses on the outside of the symphysis such that the chin represents a strengthening buttress to prevent breakage.

Riesenfeld (1969) investigated "chin" development in laboratory animals. Mandibular shortening was induced experimentally in rats by extraction of the incisors and excision of both masseter and temporalis muscles. The ensuing morphological changes closely paralleled the phylogenetic changes in the evolution of the hominid mandible: reduction of the symphyseal angle remotely approaching the condition of a chin, increase of the mandibular angle leading to a more arched alveolar border, forward movement of the lower symphyseal border, loss of inversion of the lower mandibular border with a change to eversion of the entire mandibular body so that the molar teeth can be seen below, slightly more accentuated inversion of the condyles, and frequently eversion of the greatly reduced angular process. On the basis of these results, Riesenfeld concluded that the chin resulted neither from compression of the condyles produced by the external pterygoid muscles nor from changes in the lower border of the mandible following adaptation to an orthograde posture. Rather, chin formation was considered to arise merely from a shortening of the mandible.

Obviously, therefore, until the controversy relating to the origin and function of the "chin" and its relationship with the simian shelf has been clarified, the significance of these two morphological attributes will remain conjectural. Furthermore, it is difficult to interpret the evolutionary implications of the direction and location of the mental foramen (Simonton, 1923; Montagu, 1954) until these controversies have been resolved.

Proponents of the hypothesis that heredity is the primary determinant of jaw size and shape include Brash (1929). The general form of the cartilaginous skeleton in higher vertebrates is determined by factors inherent to the rudiment. In addition, specific mechanical conditions are not necessary for the differentiation or continued activity of osteoblasts or for the initiation or activation of osteogenesis. Watnick (1972) contrasts the mandibular morphology of monozygote and dizygote twins and showed that analysis of small unit areas, representing local growth sites, reveals different modes of control within the same bone. The variability of the lingual symphysis, lateral surface of the ramus and frontal curvature of the mandible is mainly genetically determined. In contrast, variability of the gonial angle and antegonial notch is predominantly determined environmentally. Furthermore, the posterior border of the ramus, subsymphyseal contour and labial symphysis are controlled by both genetic and environmental factors. Thus, even within a relatively simple bone, the component parts are under variable genetic and environmental control.

OCCLUSOFACIAL RELATIONSHIPS

Data from standardized lateral cephalographs of caucasoid children and adults indicate that the occlusal aspect of the normal human maxillary dentition relates within the facial complex in a relatively invariant manner. For instance, the angular relationship between the occlusal plane and facial plane is characterized by small variability when judged by the size of its standard deviation as compared with those of other cephalometric angular values. Also, linear measurements reveal that the occlusal surface reference points used in the construction of the occlusal plane are themselves orientated to the nasion, so that their relative distances to this skeletal point show small absolute variability. This linear occlusofacial relationship of small variability was found in a variety of occlusal types, suggesting that it is a primary attribute determining the angular constancy of the occlusal plane to the facial plane evident in occlusal normals (Zingeser 1960).

Subsequently, Zingeser (1966) contrasted the relationship elucidated for man with the howler monkey (*Alouatta caraya*), where the data showed comparable small variability in the orientation of the maxillary occlusal region within the facial complex. The howler monkey, in common with most nonhuman primates and other mammals, ontogenetically develops a marked occlusofacial prognathism associated with mesial migration of the dentition. As a result, the relative relationship of occlusal loci to the earlier stabilized circumorbital and interorbital skeletal points change with facial growth. In contrast, the marked vertical growth vector characteristic of human facial development occurs without substantially altering the relative distances of occlusal loci to nasion. Hence, the human pattern is constant throughout a wide range of age levels (Zingeser, 1960).

These characteristics of small variability are prevalent in large populations, suggesting a probable high order of adaptive significance (Muller, 1967). The development and maintenance of occlusal precision is of prime importance to survival of an animal exposed to natural selection. This, therefore, supports the notion that the entire facial region comprises as a functional matrix, the rigid dimensional characteristics of which relate to the vital process of attaining and maintaining occlusal precision (Moss and Young, 1960).

TEMPOROMANDIBULAR JOINT

In the "generalized" mammalian form (Biegert, 1956) both dental arch and temporomandibular joints lie at the same level, e.g. in the lemur, while in the "universally specialized" type, the temporomandibular joints have ascended relative to the dental arches resulting in a superelevation of the mandibular condyles relative to the mandibular corpus. Taxonomically, the distinctive character of the temporomandibular joint is the new development of an articular eminence (tuberculum articulare). This, contrary to tradition, is not a special hominid character; neither are the small glenoid processes and posterior delimitation of the articular pit by the tympanicum (Biegert, 1956). Indeed, if the temporomandibular joints of australopithecines appear to exhibit a hominid tendency, this arises primarily from the angle subtended by the tympanicum and pars glenoidalis, which, when viewed sagittally, is more acute than in the majority of pongids. This condition arises from a greater "rolling up" of the neurocranium compared with pongids. In hominoids, the temporomandibular joints are vertically displaced from the occlusal surface of the upper dental arcade. This could have arisen from an alteration in the position of the upper jaw relative to the neurocranium by an upward tilting of the upper jaw, (a specialization occurring in lower evolutionary stages, e.g. lemuriforms). Alternatively, there may have been a vertical displacement of the nasal cavity and dental arch relative to the anterior cranial base and orbital cavities, or additional ventral displacement of the dental arch relative to the floor of the nasal cavity due to palatal vaulting. These last two conditions are found in more highly evolved forms (simians) and there also appears to be a consistent reduction in the olfactory organs together with the tendency to develop a higher subcerebrally placed upper jaw, i.e. a more vertically orientated facial skeleton.

Although the changes in the temporomandibular joint during primate evolution have received scant attention (Todd, 1930; Haussen, 1931) numerous morphological descriptions of the human and ape temporomandibular joints have been provided.

Knowles (1915) reported that the glenoid fossa of Eskimo skulls is not so deeply concave as that from "civilized races," but tends to be shallow. Furthermore, "many Australian crania have deep, well-marked glenoid fossae, but this also applies to other races." In anthropoid apes, the glenoid fossae are shallow and the articular eminence flattened; those of other hominids being intermediate between man and apes (Knowles, 1915). Discrimination is often difficult, however, between the glenoid fossae of man and anthropoids from casual inspection. Nevertheless, despite their general similarity, observation is often the only criterion used to discriminate between the glenoid fossae of man and apes. This is surprising in view of the fact that the articular fossa varies in shape between different ethnic groups of man, at different ages in an individual, and on occasion between both sides of the same individual. There are also pronounced variations in the depth of the fossa and its contours in man, although these are no greater than in anthropoid skulls. The shape of the head of the articular process of the mandible and of the articular eminence are plastic and vary both with age and mechanical functioning of the jaws.

In an attempt to overcome this deficiency, Ashton and Zuckerman (1952) measured the mastoid and digastic fossae in fossil and extant primate groups. They concluded that, in this respect, australopithecines closely resembed pongids, athough the pongids may exhibit a wide, shallow glenoid cavity with no articular eminence, whereas hominids, including australopithecines, have been described with a deep, narrow fossa bearing a marked articular eminence (Robinson, 1952). On the basis of this differentiation, Broom and Schepers (1946) and Broom, Robinson and Schepers (1950) reported that the temporomandibular joint of australopithecines is more human than ape-like, although subsequently australopithecines have been described with both types of glenoid fossae (Robinson, 1952). In view of the lack of datum points, however, it is difficult to define the glenoid fossae metrically. For example, the means for anterioposterior fossa length range between 19 to 27 mm and for pongids 29 to 46 mm (these values were determined by Tobias (1967) on the basis of data provided by Weidenreich (1943)). From the anteroposterior dimensions, Tobias (1967) concluded that the glenoid fossa of austraopithecines is more pongid than hominid—a feature which closely concurs with the depth of the glenoid fossa. It would seem, therefore, that the anteroposterior "compression" of the glenoid fossa, which is so marked a feature of hominids (Ashton and Zuckerman, 1954) is not yet apparent in australopithecines. Nevertheless, Ashton and Zuckerman (1954) have shown that the glenoid fossa in the young human skull is more compressed than it is in the ape skull of corresponding age. Furthermore, these workers concluded that the differences between the fossae of adult pongids and hominids are essentially the result from growth occurring in all directions in apes whereas in man relatively less occurs in the anteroposterior dimension of the fossa than in other directions. Obviously, however, despite the possible lack of homologous datum points, further metric data are required before the glenoid fossa can be considered to have taxonomic significance in primates.

No significant differences appear between the condylar processes of man and apes apart from the fact that the human condyle is more slender and compressed anteroposteriorly; the human coronoid notch is greater in size relative to the mandibular ramus so that the condylar process appears relatively bigger; concomitant with the anteroposterior compression of the process, the attachment area of the external pterygoid muscle to the neck of the process is more of a pit in man than apes, resulting in the neck of the human process appearing more slender. The orientation of the transverse axis of the articular head appears to be the same in apes as in man, and in both the articular surface usually extends slightly more on the anterior than posterior surface of the condylar process. Indeed, the form of the condylar process of the mandible apparently offers little value for quantitative primate taxonomy, although future studies may well invalidate this conclusion.

From examination of scratch marks on the articular facets of teeth, Mills (1955) has demonstrated that some degree of lateral excursion of the mandible is possible in virtually all primates suggesting a common pattern of mandibular activity. This must surely preclude undue reliance on the temporomandibular articulation for primate taxonomy. Nevertheless, hypertrophy of ape canines must lead to a different degree of mandibular excursion compared with modern man, i.e. a limited narrow of lateral excursion. Also, the similarity in the order of appearance and coalescence of facets of attrition in the teeth of man and apes (Zuckerman, 1958) can only indicate that man and apes share a similar pattern of mandibular activity. This implies that the presence of large canines does not preclude lateral excursion of the ape mandible and that protrusion of the mandible neutralizes any locking of mandibular movements which might otherwise result from large canines. The teeth of australopithecine fossils exhibit a similar pattern of tooth attrition compared with extant apes. Thus, these specimens probably enjoyed a similar pattern of mandibular activity. This all provides evidence to the effect that temporomandibular joint morphology has dubious taxonomic value.

Gorilla and australopithecine fossils provide two contrasting forms with large teeth, a large glenoid fossa and poorly expanded frontal region of the neurocranium. When such pongids and hominids are compared with modern man, neurocranial expansion converts the glenoid fossa from being mainly outside the sagittal plane of the lateral cranial wall to being mainly or even wholly within the lateral plane (Weidenreich, 1943). The significance of the lateral expansion of the glenoid fossa in australopithecines in contrast to the medial expansion in *Gorilla* possibly lies in the fact that the interglenoid and biglenoid distances reflect the intercondylar and bicondylar distances of the mandible. Thus, austalopithecines appear to have achieved an increase in dental, mandibular, and condylar size without a narrowing of the space between the two halves of the mandible. The functional significance of this morphological change remains obscure.

As with most skeletal structures, the morphology of the temporomandibular joint is subject to pathological changes (reviewed by Alexandersen (1967). It is, therefore, important to exclude joints exhibiting such changes when attempting a metrical definition of the temporomandibular joint surfaces of primates. Furthermore, James (1960) remarked that in macerated skulls, the articular or joint surfaces are not covered by tissue whereas, in life, there is a variable covering of articular fibroconnective tissues. Therefore, this variable covering may possibly result in a different temporomandibular joint morphology than can be demonstrated in macerated skulls and, of course, such a tissue configuration cannot be estimated from examination of fossil specimens. This may again denigrade the taxonomic significance of primate temporomandibular articulations.

DENTAL ARCH

There is scant data regarding the arrangement of tooth roots arch and dental arch proper. In man, the teeth form a continuous series without natural diastemata. This contrasts with the majority of mammals where diastemata are evident either between the incisors and canines or between the canines and premolars. Diastemata largely result from postnatal growth, i.e. from differential alveolar growth in a horizontal plane. The division of the dentition into two functionally distinct regions occurs at the beginning of mammalian evolution and is less marked in higher primates; whereas in man, it no longer exists. The presence or absence of diastemata in the forerunners of man and anthropoid apes has led to speculation as to the significance of this polymorphism for the origin of man (Gregory, 1922; Montagu, 1943; Schultz, 1948; De Boer, 1960). In anthropoid apes, diastemata are frequently present in regions associated with long-elongated canines (Schultz, 1948); whereas both canines and jaws are smaller in man. Both Gregory (1922) and Montagu (1943) have suggested that the gradual reduction in size of the primitive canines was accompanied by a reduction in jaw size and the subsequent closure of the interproximal spaces to a point where "diastemas are lost in man."

Alignment of teeth within the arch is essentially dependent upon the relationship between jaw and tooth size, crowding and spacing arising from disproportion between these two factors. If a diastema is defined as a space greater than 0.5 mm between the proximal surfaces of adjacent teeth, then diastemata exist in

108 *Evolutionary Changes to the Primate Skull and Dentition*

caucasoids, mongoloids, and negroids (Keene, 1963; Lavelle, 1970). Obviously, more data are required before the taxonomic significance of diastemata can be appreciated in primates.

Irregularity of teeth within the dental arch has been described in a variety of primates including a number of historically "old" hominid populations. For instance, the left canine from the

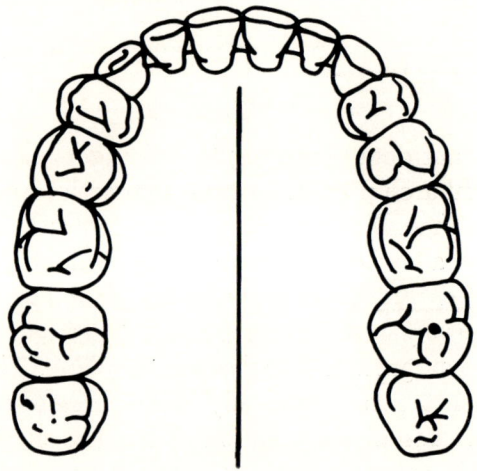

Figure 11. Upper dental arch of *Australopithecus*.

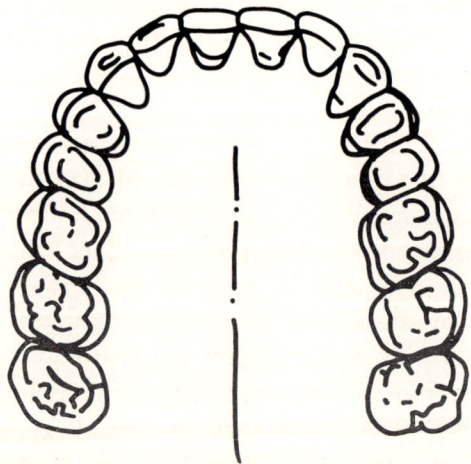

Figure 12. Upper dental arch of *Homo sapiens*.

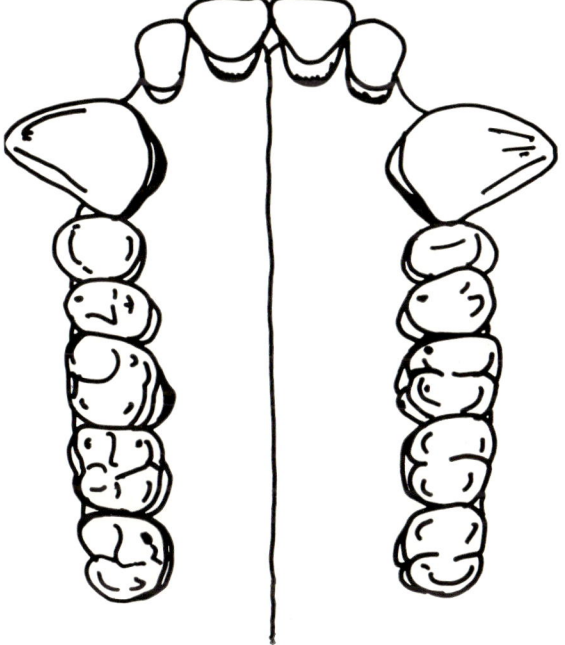

Figure 13. Upper dental arch of male *Gorilla gorilla*.

juvenile *Le Moustier* skull is impacted; whereas, the premolars are rotated in *Skhul IV* (McCown and Keith, 1939). But until a universally applicable and accepted form of categorisation of malocclusion for all primates has been established, the significance of dental overcrowding in primate taxonomy cannot be established.

The relative stability of dental arch form and the anatomic position of the arch between the tongue and circumoral musculature support the notion that the teeth reside in equilibrium in the undisturbed state. The particular dental arch form in an individual is probably a predictable segment of a larger morphologic pattern; a harmonious part of the total craniofacial architecture which is, in fact, a locally determined segment. Yet, although rational mechanisms have been proposed through which a counterbalance of intraoral forces might occur, the equilibrium concept remains essentially unverified and experimentally unproven.

Each tooth constitutes a structural unit and each exists as an individual entity subject to the physical constrictions of its oral location. A tooth is independently variable in size, shape, and mass as well as in site of origin, and positional behavior following eruption. Yet, stability of the denture reflects the balanced state of all the teeth considered together. It is the collective response of all the teeth to all environmental forces and their resultant location in the oral environment that are observed and defined as dental arch form. Arch form exists in the absence of teeth, and this leads to the notion of a biologically determined curve along and within which teeth may be considered to be transient residents.

The identification of dental arch form conventionally relies upon the presence of teeth *in situ,* for evidence of the algebraic summation and balance of all the forces to which they are subjected. In natural arch arrangements, then, the teeth themselves reveal, by the communality of their collective positions, a curve defining limits of an environmentally controlled equilibrium. Human arch form is generally described as a type of conic section curve (or catenary curve). This is a qualitative description and often refers to the dental arch as a parabola, ellipse, or segment of a circular arc of a sphere, and attempts to quantify such a form cannot withstand critical analysis.

The geometry of elliptical curves closely approximates the natural dental arch form, especially in relation to a closed elliptical curve. This concept has inherent appeal, since by regarding the dental arch as an open curve such as a parabola or caetenary, there is no known method for relating arch widths to lengths except as a parameter of tooth sizes and locations along the arch. Yet, tooth sizes are independent variables which differ even between antimeres. Tooth-dependent measurements of dental arch form reflect some estimation of variation in tooth size and locations along the curve, but are not specific parameters defining the curve itself, despite their widespread usage.

Brader (1972) investigated the pressure profile encompassing the human dental arch and concluded that the forces are essentially bisymmetric equivalents. The forces are uniformly greater in size internally to the teeth than in the buccal and labial vestibules, and diminish progressively distal to the mid-sagittal line both buccally and lingually. By superimposing the tissue pressure profile onto a scaled model of a typical human dental arch, Brader established a relationship between the pressure magnitudes and the length of the radii of curvature at the pressure sites, i.e. an inverse relationship occurred whereby the greater the pressure the shorter the radius of curvature. This finding suggested that the pattern of pressure profile relates generally with known engineering principles that obtain between pressures and the shape of any elastic container. A simplified physical model of intraoral phenomena may therefore be constructed by utilizing a rubber balloon placed inside a second balloon and having a series of loosely joined solid objects trapped between them. If the inner balloon is then inflated, the solid objects assume positions along curves, the size and form of which being determined by the pressure and tension characteristics of the two balloons. A condition of equilibrium is established along these curves where the outward pressure of the internal balloon is counterbalanced by the inward tension of the external balloon.

The hypothesis assumes that the tongue is a source of energy, and that at any instant in time, its resting potential energy is a physical constant which is expressed against the teeth from within the dental arch anteriorly and laterally from the tongue source. Another source of lingual potential energy can be accredited to resistance to change in tongue form due to its viscosity and of course the tongue's position. Brader's model assumes that the vestibular forces of the cheek and lip tissues against the teeth are inherent in the elastic tension of the circumoral tissue envelope, and that these tensions depend upon the curvilinear shape of the envelope. This conception is based on certain assumptions: the cheek-lip tissues constitute an elastic envelope, the labiolingual curve of the dental arcade is smooth in form suggesting molded determination by yielding, elastic, soft-tissue structures, the pressure outside the lips and cheeks is zero, i.e. equal atmospheric pressure on both sides, the resting tissue pressures measured labial and buccal to the teeth are internal pressures. Brader was, therefore, able to show that the teeth do reside in positions of equilibrium forced anteriorly and laterally by the internal field of forces of the tongue musculature and counterbalanced precisely along the interface between the teeth and the labiobuccal mucosa by the inward tension of the enveloping tissues. It follows, therefore, that if the force of tooth eruption results in a tooth entering the force field of the tongue in some position that encroaches upon the tongue, then in response, the potential energy of the tongue, at least partly inherent in its resistance to change in form, will influence the tooth directionally away from the tongue. The bony support of the tooth responds to the tongue pressures against the tooth crown and permits positional change of the tooth. As the teeth move away from the tongue, the radius increases and concurrently the effective tongue force diminishes in an amount sufficient to approach a state of balance with the force of the tension of the elastic envelope of the cheek. Further progress in tooth movement eventually finds the tooth encroaching into the elastic tissue of the labial or buccal envelope. In this respect, the envelope experiences a localized radius shortening, causing the force of tension of the envelope at that site to increase and thus restrict additional tooth movement into the envelope. Hence, observed dental location along the arch curve may be considered to reflect the counterbalance between energy conditions of the environmental tissues.

This concept of the forces relating to dental arch form has been formulated for man, although whether such a system is applicable to other primate groups with different arch forms is unknown. Obviously, therefore, further data are required before the functional basis of different primate dental arch forms can be established.

Although based upon datum points defined by teeth, many techniques have been used to contrast dental arch form.

The maxillary-alveolar index is based upon the maximum (or external) measurements of (a) the maxilloalveolar length—defined as the distance between the alveolar point (alveolar or prosthion) and the line linking the posterior borders of the maxillary tuberosities, and (b) the maxilloalveolar breadth—defined as the distance between the outer borders of the alveolar arch immediately above the middle of the second molar tooth. This index has been used by numerous workers to compare the maxillary dental arches between different primate samples (Table IX), and using this index, Tobias (1967) deduced that the maxil-

TABLE IX
MAXILLOALVEOLAR INDICES OF HOMINOIDS*

Specimen	Index		Worker
Zinjanthropus	95.0		Tobias, 1967
Paranthropus (SK 46)	96.0		Robinson, 1956
Australopithecus (Sts. 5)	90.0		Robinson, 1956
H. erectus erectus IV (unrestored)	104.0		Weidenreich, 1945
H. e. pekinensis (reconstructed)	107.6		
La Chapelle	101.2		
			Weidenreich, 1943
Skhul V	106.0		
Broken Hill	116.2		
Australoid (male + female; n = 134)	108.9	(mean)	Campbell, 1925
Negroid (Bantu male + female; n = 91)	110.5	(mean)	Shaw, 1931
Negroid (Bantu male; n = 334)	112.2	(mean)	
(Bantu female; n = 95)	110.3	(mean)	Jacobson, 1966
Gorilla male	64.7	(mean)	
Pongo male	74.8	(mean)	Martin, 1928
Pan male	80.8	(mean)	

Ranges of Maxilloalveolar indices of Hominoids*†

Pan	70.8– 90.4
Pongo	66.3– 85.7
Gorilla	52.1– 73.7
Australopithecinae	90.0– 96.0
Fossil *Homo*	101.2–116.2
Modern *Homo*	87.5–154.0

* Expresses as percentages.
† After Tobias, 1967.

lary arch of australopithecines is intermediate between pongids and hominids. Nevertheless, the maxilloalveolar index suffers from a number of deficiences: Only the maxillary arch is considered; the dimensions used to compute the index are imprecisely defined, such that it is difficult to compare the data from different workers; the index cannot be used for growth studies, since the maxilloalveolar breadth is measured at the level of the second molar. In addition, the index has not been extensively used, so that there is little data concerning the variation of this index either between or within different primate groups.

An alternative parameter is the palatal index, derived from palatal length and width (Petit-Maire, 1974). Palatal length is defined as the distance between the orale and staphylion, whereas palatal width is defined as the breadth between the lingual surfaces of the second molars by Olivier (1960), and the breadth between the inner aspects of the alveolar margins opposite the mesiodistal midpoints of the second molars by Montagu (1960). Table X lists the values of this index for hominoids which Tobias (1967) used to conclude that the palate of australopithecines is more ape-like than man-like. In addition, an index of 110.5 percent has been computed for Bantu negroids (Shaw, 1931) and 107.7 percent by Campbell (1925), whereas Martin (1914) lists

TABLE X
PALATAL INDICES OF HOMINOIDS*

Specimen	Index	Worker
Zinjanthropus	48.3	
H. e. pekinensis	75.1	
Skhul V	75.0	
		Tobias, 1967
La Chapelle	80.6	
Broken Hill	84.6	
Modern human races	63.6–94.6 (range)	
Pongids	34.5–62.5	Weidenreich, 1943

* Expressed as percentages

this index for other primate groups, e.g. *Papio,* 66.8 percent; *Hylobates,* 76.8 percent; male *Gorilla,* 64.7 percent; and male *Pan,* 80.8 percent. Hence, this index has taxonomic potential, although the definition of the datum points precludes its utilization for growth studies and there is again no data relating to the variation both between and within different primate groups.

The dental arch index provides an alternative quantitative measure of dental arch form, although different methods have been utilized to compute the index. For instance, in the odontometric method, dental arch length is defined as the distance between the tangent to the labial surfaces of the median incisors and a plane through the arch breadth line (first molar to first molar) perpendicular to the occlusal surface, and dental arch breadth is defined as the distance between the intersection of the main occlusal grooves of the first molars (Moorrees, 1957). In the anthropometric method, the dimensions may be defined as follows: dental arch length as the distance between the tangent to the labial surfaces of the central incisors and a plane tangent

TABLE XI

AGE CHANGES IN DENTAL ARCH AREAS IN DIFFERENT ETHNIC GROUPS

	Age (Years)	Caucasoid Male x	SE	Caucasoid Female x	SE	Negroid Male x	SE	Negroid Female x	SE	Mongoloid Male x	SE	Mongoloid Female x	SE
Maxilla	3	4.59	0.06	4.87	0.10	5.05	0.09	4.99	0.25	4.41	0.18	4.27	0.34
	4	4.65	0.06	4.85	0.05	5.21	0.11	5.01	0.06	4.46	0.30	4.30	0.21
	5	4.79	0.06	4.87	0.08	5.36	0.06	5.20	0.09	4.94	0.14	4.84	0.16
	6	6.84	0.04	7.02	0.16	7.24	0.04	7.04	0.14	5.88	0.19	5.66	0.19
	7	8.80	0.12	8.70	0.09	9.41	0.13	9.23	0.08	7.29	0.08	7.08	0.28
	8	9.27	0.02	8.25	0.07	9.85	0.04	9.35	0.12	8.31	0.14	7.95	0.30
	9	9.43	0.05	9.63	0.14	10.72	0.07	10.41	0.19	8.86	0.18	8.41	0.31
	10	9.98	0.10	9.68	0.08	13.40	0.12	13.00	0.11	9.41	0.20	9.00	0.26
	11	10.43	0.10	9.70	0.13	15.20	0.08	14.96	0.09	11.20	0.11	10.69	0.24
	12	12.04	0.04	11.83	0.07	15.21	0.11	14.98	0.16	11.38	0.07	10.72	0.11
	13	14.30	0.11	13.08	0.11	15.26	0.11	15.07	0.13	11.39	0.04	10.85	0.31
	14	14.37	0.09	13.41	0.14	15.28	0.06	15.09	0.08	11.46	0.09	10.89	0.08
	15	14.43	0.11	13.71	0.11	15.33	0.09	15.14	0.04	11.50	0.11	10.94	0.06
Mandible	3	4.49	0.06	3.42	0.07	5.06	0.18	5.00	0.14	4.44	0.16	4.37	0.04
	4	4.52	0.03	3.38	0.16	5.21	0.04	5.01	0.19	4.46	0.09	4.39	0.12
	5	4.59	0.08	3.33	0.07	6.34	0.09	6.04	0.12	5.22	0.14	5.01	0.09
	6	6.35	0.05	6.71	0.09	7.36	0.02	7.18	0.06	6.41	0.19	6.11	0.14
	7	8.41	0.17	8.20	0.09	8.52	0.06	8.44	0.09	8.28	0.08	8.14	0.18
	8	8.76	0.13	8.39	0.14	8.95	0.11	8.85	0.14	8.42	0.22	8.28	0.20
	9	9.03	0.10	8.58	0.10	9.47	0.07	9.26	0.09	9.06	0.16	8.96	0.20
	10	9.84	0.10	8.72	0.03	9.99	0.24	9.49	0.12	9.97	0.19	9.48	0.14
	11	10.13	0.14	8.83	0.11	11.42	0.08	11.24	0.06	10.28	0.25	10.11	0.34
	12	11.65	0.08	11.47	0.14	12.38	0.06	12.06	0.07	11.41	0.29	11.01	0.19
	13	12.79	0.12	12.27	0.13	12.39	0.21	12.08	0.13	12.84	0.14	12.22	0.08
	14	12.85	0.08	12.33	0.09	12.42	0.14	12.14	0.06	12.87	0.11	12.31	0.17
	15	12.87	0.06	12.38	0.06	12.54	0.19	12.20	0.11	12.88	0.11	12.36	0.16

x = mean arch area (cm^2)
SE = standard error; 20 in each age group and sex for each sample

to the distal surfaces of the second molars, perpendicular to the occlusal plane, and dental arch breadth as the greatest distance between the buccal surfaces of the second molars (Moorrees, 1957).

Finally, Laing (1955) devised a length-breadth index of the dental arcade, termed the *arcadal index.* The maximum arcadal length is defined as the distance between the midpoint of a line joining the most anterior points on the tips of the incisor teeth to the midpoint of a line joining the most posterior points of the third molar teeth, whereas the maximum breadth is defined from the buccal surfaces of the second molar teeth.

Of the three techniques for determining a dental arch index, the odontometric method uses the first molar, the anthropometric method uses the second molar and the arcadal index uses the third molar for defining the posterior extremity of dental arch length. Using the anthropometric method, the dental arch index for australopithecines lies within the range computed for pongids incorporating the internal dimensions for arch breadth, whereas using the external dimensions of arch breadth, the index for austrapolithecines lies between that computed for pongids and hominids (Tobias, 1967). Finally, with Laing's arcadal index (Table XII), the values for australopithecines lie within the lower limits of the hominid range (Tobias, 1967). All three dental arch indices have certain reservations. As with the maxilloalveolar and palatal indices, the dental arch indices have been computed for the maxillary arch only, and with all three indices, the defini-

TABLE XII
ARCADAL INDICES OF HOMINOIDS*

Specimen	Index	Worker
Zinjanthropus	97.6	Tobias, 1967
Negroids (24 Bushmen)	100–136 (range)	Laing, 1955
Talgai cranium	95	
Grimaldi youth	100	
Tasmanian	106	
Australian Aboriginal	110	
		Tobias, 1967
Predomosti	110	
La Chapelle	116	
Broken Hill and Predmosti (female)	120	
Modern English crania	120 (mean)	

* Expressed as percentages. After Tobias, 1967.

tion of the various dimensions are based upon variable datum points, e.g. the definition of dental arch length does not take into account asymmetry of the dental arch or variation in tooth position. Also, as the dimensions of arch length are based on either the first, second or third permanent molar, they cannot be utilized to examine growth changes.

In order to examine the dental arch index for both the maxillary and mandibular arches at different age groups, another arch index has been devised wherein arch width is defined as the distance between the occlusal tips of the canines (permanent or deciduous) on each side of the arch, whereas dental arch length is defined as the minimum distance between the centers of the central incisors and second premolar teeth (or their deciduous predecessors). As summarized in Table IV, such dental arch indices show variation not only between different age groups but also between different ethnic groups (Lavelle, 1970), although such analyses have yet to be applied to other primate groups, e.g. pongids. Nevertheless, Table IV emphasizes that the arch dimensions may still change after the eruption of the permanent teeth—a source of variation which must be considered in taxonomic studies.

Dental arch form has also been assessed from computation of the area enclosed within the dental arcade. For example, the area of the maxillary arch has been computed for *Pan* (3650 mm^2), Bantu negroids (2850 mm^2) and modern English caucasoids (2600 mm^2), whereas the area of the mandibular arch has been computed for the Bantu negroids (3190 mm^2), modern English caucasoids (2760 mm^2) and the Heidelberg mandible (3540 mm^2) (Keith, 1916). Moreover, from computation of maxillary and mandibular arch areas at different age groups, Lavelle (1970) has shown there to be a marked change in dental arch shape in British caucasoids during the transition periods between the deciduous and permanent dentitions. This, therefore, emphasizes the importance of variation due to age changes which must be considered in taxonomic studies (Table XIII).

It is obvious, however, that much more data must be accumulated before dental arch form can be exploited in taxonomic studies. Rather than relying on the upper dental arch, other workers

TABLE XIII
AREA OF ADULT DENTAL ARCHES OF MAN AND APES

Specimen	Maxillary arch				Mandibular arch			
	Male		Female		Male		Female	
	x	SE	x	SE	x	SE	x	SE
Australoid (Australian aborigine)	17.76	0.42	16.47	0.22	15.94	0.44	15.18	0.29
Mongoloid (N. Am. Indian)	13.22	0.39	12.22	0.17	13.01	0.09	12.41	0.36
Negroid (W. African)	18.52	0.55	17.29	0.39	15.07	0.96	14.66	0.14
Caucasoid (Anglo-Saxon)	19.99	0.44	18.31	0.46	15.21	0.85	14.94	0.37
(Moorfields — 16-18th Cent)	19.01	0.28	17.66	0.41	14.88	0.09	14.28	0.18
(St. Brides — 19th Cent)	14.48	0.34	14.25	0.42	13.68	0.11	13.19	0.19
Gorilla	48.11	0.86	24.75	0.97	40.74	0.85	22.59	0.79
Pongo	38.18	0.94	24.75	0.86	34.97	0.72	21.17	0.67
Pan	27.09	0.88	22.64	0.81	25.12	0.88	21.94	0.89

x = mean dental arch area (cm^2);   SE = standard error;   20 specimens in each sample have confined their attention to the lower dental arch. For instance, the arrangement of the teeth in the mandible of man may be distinguished from that of apes by means of the "basic rectangle of the mandible." This rectangle has been defined by Leakey (1968) by drawing a line tangential to the anterior alveolar margins of the two central incisor teeth and another tangential to the two most posterior points on the third molars. When this had been done, perpendiculars were dropped from the two most posterior points of the third molars to meet the line tangential to the incisors. Naturally, there is some degree of individual variation within any genus or species. Nevertheless, Leakey (1968) was able to show that despite these minor variations, the shape of the basic rectangle is remarkably constant for modern Pongidae on the one hand and modern man on the other. In Pongidae, the sides of the rectangle are almost twice as long as the ends, whereas in both *Homo sapiens* and *Homo erectus*, the sides of the rectangle are virtually of the same length. Leakey (1968) presented several illustrative diagrams, but otherwise no attempt was made to quantify this difference.

The basic rectangle of the mandible may be treated more precisely by computing an index derived from measurements of the rectangle. When the anteroposterior length of the rectangle is divided by its breadth and expressed as a percentage, an index results termed the "basic rectangle of the mandible index (BRM index)." Kinzey (1970) used this index to compare nineteen genera of living primates and noted striking racial differences between American caucasoids (86.7 ± 2.2) and American negroids (96.6 ± 1.2). In addition, this index revealed considerable contrasts between other primate groups, e.g. 155.6 ± 2.1 for *Pan*, 172.9 ± 2.2 for *Gorilla*, 174.7 ± 3.7 for *Pongo*, and 138.2 ± 1.8 for *Hylobates*. This index is not particularly sensitive to morphological differences however, since sexual dimorphism appeared to be insignificant for apes and man although there are considerable contrasts between the dental arches of male and female man and apes (Morant, 1936; Giles, 1964). Furthermore, from computation of the BRM index, Kinzey (1970) noted that the mandibular dental arch of australopithecines was within the range of living pongids, whereas *Kenyapithecus africanus* and *Dryo-*

pithecus africanus were within the ranges of American negroids.

In computation of the BRM index, Kinzey (1970) included the third molar tooth as a datum point which is notoriously variable in man and apes. Hence, another study was undertaken to compare the BRM indices between different primate groups, extending the basic rectangle only to the mandibular first permanent molar tooth.

The results shown in Table XIV show that the BRM index does not significantly increase or decrease with age in the three ethnic groups. Furthermore, although the BRM indices were similar for both caucasoids and negroids, they were considerably lower at all ages for mongoloids, due to the basic rectangle being wider in the latter compared with negroids and caucasoids.

In the second part of this study, the BRM index was determined for the skulls of different ethnic groups of man and other primate groups. As shown in Table XV, the BRM index was similar for negroids and caucasoids, and similar although lower for mongoloids and australoids. Furthermore, the BRM indices were consistently higher for the other primate groups compared with man, indicative of a longer and narrower mandibular arch for the former compared with the latter. Thus, it would appear that this index could provide significant information to be used in primate taxonomy.

In order to test this hypothesis, the two dimensions of the basic rectangle were determined for the mandible of three neanderthal specimens, two from *Australopithecus africanus* and two from *Australopithecus robustus*. The BRM index for neanderthals averaged 79.0 overlapping the range for modern man, whereas that for *Australopithecus africanus* averaged 107.2 and that for *Australopithecus robustus* averaged 111.3—both lying between the ranges of the BRM index for man and apes. There are, however, a number of factors which influence the shape of the basic rectangle:—the degree of divergence of the teeth rows, the location of the bend towards the symphysis, the length of the teeth rows, and the degree of prognathism of the anterior alveolar region. Moreover, although Kinzey (1970) concluded that on the basis of all four factors, the basic rectangle could be used to separate recent hominids from recent pongids, only the location of

TABLE XIV
AGE CHANGES IN DENTAL ARCH INDEX IN DIFFERENT ETHNIC GROUPS

	Age (Years)	Caucasoid Male x	SE	Caucasoid Female x	SE	Negroid Male x	SE	Negroid Female x	SE	Mongoloid Male x	SE	Mongoloid Female x	SE
Maxilla	3	79.4	0.35	78.9	0.76	85.4	0.49	84.4	0.32	63.8	0.39	60.1	0.27
	4	78.7	0.28	78.8	0.65	85.9	0.38	85.1	0.49	65.1	0.72	61.1	0.11
	5	79.4	0.31	78.2	0.62	86.5	0.21	86.0	0.36	68.4	0.31	61.4	0.18
	6	79.7	0.26	79.4	0.59	87.2	0.14	86.3	0.48	70.9	0.30	62.7	0.30
	7	81.8	0.38	80.4	0.46	87.8	0.11	87.7	0.33	72.3	0.26	62.1	0.32
	8	84.0	0.42	82.1	0.43	88.6	0.10	88.1	0.85	74.0	0.34	62.5	0.26
	9	86.1	0.56	83.3	0.32	89.7	0.86	90.9	0.91	74.2	0.11	62.0	0.34
	10	88.2	0.39	86.1	0.38	90.1	0.27	90.0	0.48	75.6	0.18	62.9	0.28
	11	89.2	0.45	88.2	0.47	90.4	0.15	90.3	0.36	76.4	0.14	64.5	0.36
	12	88.2	0.47	88.2	0.49	91.0	0.36	90.7	0.37	77.1	0.29	64.0	0.34
	13	88.9	0.38	88.0	0.52	93.2	0.38	91.1	0.14	77.8	0.11	64.4	0.09
	14	87.4	0.46	87.8	0.56	93.6	0.41	91.4	0.11	77.4	0.17	64.7	0.07
	15	87.4	0.39	87.8	0.47	93.6	0.48	91.7	0.09	77.4	0.19	64.8	0.11
Mandible	3	86.3	0.57	86.0	0.68	104.3	0.44	93.0	0.24	72.3	0.22	68.4	0.15
	4	87.9	0.28	87.6	0.61	105.9	0.21	95.7	0.52	73.8	0.14	70.0	0.21
	5	87.9	0.31	88.3	0.17	106.7	0.49	97.9	0.18	76.3	0.28	72.6	0.18
	6	89.8	0.16	88.2	0.29	107.0	0.38	97.1	0.39	79.7	0.30	72.0	0.22
	7	92.5	0.29	90.3	0.38	109.2	0.37	97.1	0.11	80.1	0.32	72.8	0.26
	8	95.7	0.85	90.6	0.35	111.8	0.26	98.4	0.18	81.1	0.26	73.5	0.34
	9	95.7	0.88	91.4	0.29	113.3	0.14	98.8	0.26	82.9	0.39	73.1	0.14
	10	99.2	0.67	93.3	0.17	114.0	0.95	99.6	0.37	83.5	0.40	73.2	0.29
	11	102.4	0.45	96.3	0.65	115.0	0.86	99.7	0.28	83.1	0.32	73.9	0.14
	12	104.3	0.44	102.9	0.68	116.6	0.24	99.6	0.32	83.8	0.38	73.7	0.19
	13	106.1	0.52	102.9	0.59	117.4	0.46	99.4	0.36	83.4	0.41	74.8	0.27
	14	106.0	0.39	103.9	0.47	118.9	0.49	99.8	0.38	83.3	0.34	74.4	0.18
	15	107.6	0.31	104.2	0.28	120.1	0.35	99.7	0.41	83.3	0.32	74.9	0.14

Arch Index = $\dfrac{\text{Width between canines}}{\text{Length between central incisor and second premolar (or deciduous second molar)}} \times 100$;

x = mean index (percent); SE = standard error; 20 in each age group and sex for each sample.

TABLE XV
INDEX OF BASIC RECTANGLE IN MATURE SKULLS

	Sex	No. in Sample	Mean	Standard Error
Caucasoid (Anglo-Saxon)	M	20	96.35	1.46
	F	20	91.49	3.21
Negroid (W. African)	M	20	96.63	1.39
	F	20	92.21	2.41
Mongoloid (N. Am. Indian)	M	20	86.47	2.67
	F	20	86.14	1.04
Australoid (Australian Aboriginal)	M	20	86.95	3.46
	F	20	85.41	1.97
Gorilla	M	20	131.3	2.06
	F	20	139.3	3.47
Pan	M	20	126.3	1.49
	F	20	123.9	2.67
Pongo	M	20	140.9	1.85
	F	20	133.3	1.06
Hylobates	M	20	110.5	1.94
	F	20	109.6	2.18
C. aethiops	M	20	134.5	1.52
	F	20	131.6	2.44
Macaca	M	20	147.0	1.87
	F	20	143.8	1.89
Colobus	M	20	124.5	3.24
	F	20	121.6	2.76
Papio	M	20	170.7	1.85
	F	20	163.0	2.49
H. sap Neanderthalensis	Tabum I		77.8	
	Spy I		77.8	
	Le Mouster		81.4	
		Mean	79.0	
A. africanus			106.2	
			108.2	
		Mean	107.2	
A. robust			110.9	
			111.6	
		Mean	111.3	

the turn towards the mandibular symphysis could be used to separate Miocene hominids from Miocene apes. On the basis of this, therefore, Kinsey (1970) considered that the use of the basic rectangle was limited in Miocene hominoids.

In contrast to studies relating to the human dental arch, little data are available concerning the variation in nonhuman primate arches. The arch dimensions between different ethnic groups have been listed by Lavelle, Flinn and Foster (1971) and for *Pan* by Schultz (1940), *Gorilla* (Bolk 1925; Randall, 1943-44), and *Macaca* (Baume and Becks, 1950). In primate taxonomic studies, however, the dental arch has been generally neglected, the teeth alone being the principal organs of study. Therefore, it is possibly for this reason, that there is controversy concerning the data relating to the dental arches of fossil specimens. For instance, the arches of *Zinjanthropus* are considered to be similar to those of other australopithecines and hominids presenting an evenly arched parabolic curve (Clark, 1955; Robinson, 1956), but differ markedly from the typical pongid pattern where the canines premolars and molars follow two parallel lines or rows separated by a distinct diastema from the anterior teeth (Clark, 1952). In contrast, Tobias (1967) considers the dental arches of *Australopithecus* and *Zinjanthropus* to be relatively more narrow and longer than those for hominids. Such controversies can only be resolved, however, by the accumulation of more metric data concerning the dental arches of different primate groups.

Also using multivariate statistical techniques based on many arch dimensions, similar growth patterns for *Pan, Gorilla,* and *Pongo* have been described (Lavelle and Flinn, 1972), and this closely accords with the growth patterns for the skull and facial skeleton (Krogman, 1931) and endocranial capacity (Ashton and Spence, 1958). To compare the arch growth patterns between man and other primate groups, however, necessitates some homologous method for the assessment of age, which is not possible in view of the variation in tooth eruption sequences.

Some degree of heritability affecting dental arch form is apparent from family (Bowden and Goose, 1968) and twin (Lundstrom, 1948) studies. Also apart from exhibiting variable degrees of asymmetry (Minkin, 1925), the shape of the dental arch depends on the form and position of the tooth crowns (Scott, 1967). Nevertheless, until more data are available relating to the factors affecting dental arch size and shape, care must be taken in utilizing dental arch form for taxonomic studies.

As with many other craniofacial features, much data are required before the evolutionary changes to the primate masticatory system can be elucidated. It is indeed surprising that, in view of the preferential fossilization of the jaws, more data are not available relating to the primate jaws, particularly the upper jaw.

CONCLUSION

It is generally considered that the evolutionary changes to the primate brain started earlier in phylogeny than in other mammalian phyla. An influence of the evolving part of the brain is evident only in those phylogenetically early genera in which a specialization of the eyes for nocturnal life leads to a tremendous enlargement of the area striata and incidently to a corresponding enlargement of the occipital region of the skull. In the skull of *Tarsius*, the shape of the brain and neurocranium is also influenced by the enormous size of the eyes, each eyeball having approximately the same volume as the brain itself. Thus, the whole of the skull is influenced by the eyes and the visual system. The breadth of the occipital region also leads in part to the wide separation of the temporomandibular articulations, which in turn influences the angle of convergence of the mandibles. If the mandibles do not develop a broad, almost transversely oriented mental region, but join each other at an angle, then the more this angle will be affected by the distance between the two temporomandibular articulations, the shorter the facial part of the cranium and the mandible. The shortness of the mandible is associated with the shortness of the facial part of the skull in primates. Furthermore, the position of the premolar and molar rows does not necessarily correspond to the angle of the mandibles. It does so only when the position direction of the alveolar process corresponds with that of the corpus of the mandible. The discrimination between the position of the mandibular corpus and alveolar process is important, since only the position of the mandibular corpora to the midsaggital plane is influenced by the location of the temporomandibular articulations. Moreover, there seems to be a connection between the development of the eyes, the occipital lobes, the temporal lobes, the width of the occipital region of the skull, and the convergence of the mandibles, so that the brain comprises merely a link in the chain of mutual relation of the head organs. This, therefore, emphasizes the fact that future studies of primate jaws must be integrated with the skull as a biological and functional entity, rather than attempting to examine the evolutionary changes of the jaws in isolation.

REFERENCES

Alexandersen, V.: The pathology of the jaws and the temporomandibular joint. In Brothwell, D. R. and Sandison, A. T. (Eds.): *Diseases in Antiquity*. Springfield, Thomas, 1967, pp. 551-595.

Ashton, E. H. and Zuckerman, S.: Age changes in the position of the occipital condyles in the chimpanzee and gorilla. *Am J Phys Anthropol*, 10:277, 1952.

Ashton, E. H. and Zuckerman, S.: The anatomy of the articular fossa (fossa mandibularis) in man and apes. *Am J Phys Anthropol*, 12:29, 1954.

Avis, V.: The significance of the angle of the mandible. *Am J Phys Anthropol*, 19:55, 1961.

Barber, C. G., Green, L. J., and Cox, G. J.: Effects of the physical consistency of the diet on the condylar growth of the rat mandible. *J Dent Res*, 42:848, 1963.

Baume, L. J.: The postnatal growth of the mandible of the *Macaca mulatta*. *Am J Orthod*, 39:228, 1953.

Baume, L. J. and Becks, H.: Development of the dentition of *Macaca mulatta*. *Am J Orthod*, 36:723, 1950.

Biegert, J.: Das Kiefergelenk der Primaten, seine Altersveranderungen und Spezialisationen in Gestaltung und Lage. *Morph Jb*, 97:249, 1956.

Bilsborough, A.: Evolutionary change in the hominoid maxilla. *Man*, 6:473, 1971.

Bilsborough, A.: Cranial morphology of neanderthal man. *Nature* (Lond), 237:351, 1972.

Bjork, A.: Variation in the growth pattern of the human mandible: longitudinal radiographic study by the implant method. *J Dent Res*, 42:400, 1963.

Bolk, L.: *Die Entstehung des Menschenkinnes*. Amsterdam, Verhandl d. Koninhl. Akad. Wetensch, 1924.

Bolk, L.: On the existence of a dolichocephalic race of *Gorilla*. *Proc K Akad Wet*, 28:204, 1925.

Boyd, T. G., Castelli, W. A. and Huelke, D. F.: Removal of the temporalis muscle from its origin. *J Dent Res*, 46:1064, 1967.

Brader, A. C.: Dental arch form related with intra-oral forces. *Am J Orthod*, 61:541, 1972.

Brash, J. C.: *The Aetiology of Irregularity of the Teeth*. London, Dental Board of the United Kingdom, 1929.

Brodie, A. G.: On the growth of the jaws and the eruption of the teeth. *Angle Orthod*, 12:109, 1942.

Broom, R. and Robinson, J. T.: Swartkrans ape-man, *Paranthropus crassidens*. *Transv Mus Mem*, No. 6, 1952.

Broom, R., Robinson, J. T., and Schepers, G. W. H.: Sterkfontein ape-man, *Pleisianthropus*. *Transv Mus Mem*, No. 4, 1950.

Broom, R. and Schepers, G. W. H.: The South African fossil ape-man, the Australopithecinae. *Transv Mus Mem,* No. 2, 1946.
Campbell, T. D.: *Dentition and Palate of the Australian Aboriginal.* Adelaide, Hassel Press, 1925.
Clark, W. E. Le Gros.: Hominid characters of the australopithecine dentition. *J R Anthrop Inst,* 80:37, 1952.
Clark, W. E. Le Gros.: *The Fossil Evidence for Human Evolution.* Chicago, Chicago U P, 1955.
Cleaver, F. H.: A contribution to the biometric study of the human mandible. *Biometrika,* 29:80, 1937.
Crompton, A. W.: On the lower jaw of the *Diarthrognathous* and the origin of the mammalian lower jaw. *Proc Zool Soc* (Lond), 140:697, 1963.
De Boer, J. G.: Diastemas. *T Tandheelk,* 67:87, 1960.
Du Brul, E. L. and Sicher, H.: *The Adaptive Chin.* Springfield, Thomas, 1954.
Enlow, D. H.: *The Human Face.* New York, Har-Row, 1968.
Enlow, D. H.: Stress analysis of the facial skeleton of *Gorilla* by means of the wire strain gauge method. *Primates,* 14:37, 1973.
Enlow, D. H. and Bang, S.: Growth and remodelling of the human maxilla. *Am J Orthod,* 51:446, 1965.
Enlow, D. H. and Harris, D. B.: A study of the postnatal growth of the human mandible. *Am J Orthod,* 50:25, 1964.
Giles, E.: Sex determination of discriminant function analysis of the mandible. *Am J Phys Anthropol,* 22:129, 1964.
Gregory, W. K.: *The Origin and Evolution of the Human Dentition.* Baltimore, Williams and Wilkins, 1922.
Harris, J. E. and Kowalski, C. J.: All in the family. *Am J Orthod,* 69:493, 1976.
Harrower, G.: A biometric study of one hundred and ten Asciatic mandibles. *Biometrika,* 20B:279, 1928.
Hershkovitz, P.: Primate chin. *Bull Field Mus Nat Hist,* 41:6, 1970.
Hooton, E. A.: The evolution and devolution of the human face. *Am J Orthod,* 32:657, 1946.
Hrdlicka, A.: Lower jaw. *Am J Phys Anthropol,* 27:281, 1940.
Humphrey, G. M.: Results of experiments on the growth of the jaws. *Br J Dent Sci,* 6:550, 1863.
Hunter, J.: *Experiments and Observations on the Growth of Bones.* London, Longman, 1837.
James, W. W.: *The Jaws and Teeth of Primates.* London, Pitman, 1960.
Keene, H. J.: Distribution of diastemas in the dentition of man. *Am J Phys Anthropol,* 21:437, 1963.
Keith, A.: *The Antiquity of Man.* London, Williams and Norgate, 1916.
Kinzey, W. G.: Basic rectangle of the mandible. *Nature* (Lond), 228:289, 1970.

Knowles, F. H. S.: The glenoid fossa in the skull of the Eskimo. *Geol Survey Mus Bull,* No. 9 Anthrop ser No. 4, 1915.
Krogman, W. M.: Studies in growth changes in the skull and face of anthropoids. *Am J Anat,* 47:89, 325, 343, 1931.
Kurisu, K., Kato, S., and Kanda, S.: A factor analysis of the environmental correlation matrix from the craniofacial traits of male monozygotic twins in growing age. *Med J Osaka Univ,* 22:207, 1972.
Laffont, J. and Aaron, C.: Etude comparee differentes techniques de mesure du prognathisme. *Bull Mem Soc Anthrop* (Paris), 2:382, 1961.
Laing, J. D.: The aroadal index. *J Dent Assoc S Afr,* 10:376, 1955.
Lavelle, C L. B.: The distribution of diastemas in different human population samples. *Scand J Dent Res,* 78:530, 1970.
Lavelle, C. L. B.: A comparison between the mandibles of Romano-British and nineteenth century periods. *Am J Phys Anthropol,* 36:213, 1972.
Lavelle, C. L. B., and Flinn, R. M.: Dental arch growth in the gorilla, chimpanzee and orang-utan. *Archs Oral Biol,* 17:1095, 1972.
Lavelle, C. L. B., Flinn, R. M. and Foster, T. D.: Dental arches in various ethnic groups. *Angle Orthod,* 41:293, 1971.
Leakey, L. S. B.: Upper Miocene primates from Kenya. *Nature* (Lond), 218:527, 1968.
Lysell, L.: A biometrical study of occlusion and dental arches in a series of mediaeval skulls from Northern Sweden. *Acta Odont Scand,* 16:177, 1958.
McGown, T. D. and Keith, A.: *The Stone Age of Mount Carmel.* Oxford, Clarendon Press, 1939.
Martin, R.: *Lehrbuch der Anthropologie.* Jena, Gustav Fischer, 1914.
Martin, E. S.: A study of an Egyptian series of mandibles, with special reference to mathematical methods of sexing. *Biometrika,* 28:149, 1936.
Moorrees, C. F. A.: *The Aleut Dentition.* Cambridge, Harvard U Pr.
Montagu, M. F. A.: Variation of the diastemata in the dentition of the anthropoid apes and its significance for the origin of man. *Am J Phys Anthropol,* 1:325, 1943.
Montagu, M. F. A.: The direction and position of the mental foramen in the great apes and man. *Am J Phys Anthropol,* 12:503, 1954.
Montagu, M. F. A.: *A Handbook of Anthropometry.* Springfield, Thomas, 1960.
Morant, G. C.: A biometric study of the human mandible. *Biometrika,* 28:84, 1936.
Moss, M. L.: Functional analysis of the human mandibular growth. *J Prosth Dent,* 10:1149, 1960.
Moss, M. L. and Salentijn, L.: The primary role of the functional matrices in facial growth. *Am J Orthod,* 5:566, 1969.
Moss, M. L. and Young, R. W.: A functional approach to craniology. *Am J Phys Anthropol,* 18:281, 1960.

Muller, G. M.: Morphologische Variationen am Gesichtsschadel erbgleicher Zwillinge. *Deutsch Zahnaerztl Z, 22*:914, 1967.

Murphy, T.: The chin region of the Australian aboriginal mandible. *Am J Phys Anthropol, 15*:517, 1957.

Olivier, G.: *Pratique Anthropologique*. Paris, Vigot Freres, 1960.

Petit-Maire, N.: Anterior alveolar arch reduction and hominid evolution, with special reference to the position of *Ramapithecus. J Human Evol, 3*:237, 1974.

Randall, F. E.: The skeletal and dental development and variability of the gorilla. *Hum Biol, 15*:236, 307, *16*:23, 1943-1944.

Riesenfeld, A.: The adaptive mandibles: An experimental study. *Acta Anat, 72*:246, 1969.

Ringqvist, M.: Isometric biteforce and its relation to dimensions of the facial skeleton. *Acta Odont Scand, 31*:1, 1973.

Robinson, J. T.: The australopithecines and their evolutionary significance. *Proc Linn Soc* (Lond), *3*:196, 1952.

Robinson, J. T.: The dentition of the *Austral epithecinae. Transv Mus Mem*, No. 9, 1956.

Robinson, L.: The story of the chin. *Rep Smithson Inst, 1914*:599, 1914.

Schultz, A. H.: Growth and development of the chimpanzee. *Contrib Embryol Carneg Inst, 28*:1, 1940.

Schultz, A. H. The relation in size between premaxilla, diastema and canine. *Am J Phys Anthropol, 6*:163, 1948.

Scott, J. H.: Muscle growth and function in relation to skeletal morphology. *Am J Phys Anthropol, 15*:197, 1957.

Simonton, F. V.: Mental foramen in the anthropoids and in man. *Am J Phys Anthropol, 6*:413, 1923.

Tappen, N.: A comparative functional analysis of primate skulls by the split-line technique. *Hum Biol, 26*:210, 1954.

Tobias, P. V.: *Olduvai Gorge*. London, Cambridge U Pr, 1967.

Virchow, H.: Muskelmarken am Schadel. *Z Ethnol, 42*:638, 1910.

Watnick, S. S.: Inheritance of craniofacial morphology. *Angle Orthod, 42*: 339, 1972.

Watt, D. G. and Williams, C. H. M.: The effects of the physical consistency of the food on the growth and development of the mandible and maxilla of the rat. *Am J Orthod, 37*:895, 1951.

Wegener, K.: Ulber Zweck und Ursache der menschlichen Kinnbildung. *Z Morph Anthrop, 26*:165, 1927.

Weidenreich, F.: Das Menschenkinn und seine Entstehung. *Erg Anat Entw Gesch, 31*:1, 1934.

Weidenreich, F.: Observations on the form and proportions of the endocranial casts of *Sinanthropus pekinensis*, other hominids and the great apes. *Palaeont Sin*, D7:1, 1936.

Weidenreich, F.: The skull of *Sinanthropus pekinensis*. *Palaeont Sin, 127*: 1, 1943.

Zingeser, M. R.: Nasomaxillary proportional constancy. *Am J Orthod, 46*: 674, 1960.

Zingeser, M. R.: Occlusofacial relationships in the mature howler monkey (*Alouatta caraya*). *Am J Phys Anthropol, 24*:171, 1966.

Zuckerman, S.: Correlation of change in evolution of higher primates. In Huxley, J., Hardy, A. C., and Ford, E. B. (Eds.): *Evolution as a Process*. London, Allen and Unwin, 1958.

CHAPTER FOUR
THE PRIMATE DENTITION
INTRODUCTION

IN UNSPECIALIZED mammalian dentitions, incisors serve chiefly for cutting, canines for grasping and piercing, and premolars and molars for chewing. In many mammalian orders, not least in primates, the dentition has become adapted for a variety of other purposes such as grooming, holding, and manipulative procedures. Most of these secondary functions involve the anterior rather than posterior teeth, the latter generally remaining predominantly organs of mastication, morphologically well adapted to withstand occlusal wear. Consideration of teeth purely as masticatory organs is thus an oversimplification. This is especially true in modern man where, in addition to their use in a wide variety of cultural activities, teeth have a considerable aesthetic and sexual appeal which may well have been and still be of selective significance. While Osborn (1973) has reviewed the evolution of the primitive Eutherian dentition, only the primate dentition will be discussed in this chapter.

CONTROL OF TOOTH DEVELOPMENT

The orderly sequence of morphological and biochemical events culminating in tooth formation are products of tissue interactions occurring during development. Kollar and Baird (1969) have demonstrated that isolation of the epithelial and mesenchymal components prevents further development in each component. From subsequent investigations of Kollar and Baird (1971), it appears that the structural specificity for three-dimensional tooth shape resides in the mesoderm. In order to gain insight into the way this structural specificity is controlled, it is necessary to describe a number of studies which were published long before this modern work.

Genetic Control of Tooth Form

In 1894, Bateson suggested that the teeth could be likened to structures such as vertebrae and digits, in that the structure is serially repeated within the body of the individual. This phenomenon is termed "Merism." Bateson went on to describe some of the properties of meristic series as follows: (a) all members of the series are constructed to a common plan, they show serial homology and the differences between them are quantitative rather than qualitative; (b) the series shows internal differentiation, the character of each member depending on its position in the series as a whole; (c) individual members of the series can vary so as to acquire characters proper to adjacent members. It is clear that the dentition has these properties. Bateson regarded a meristic series as a unity, varying and evolving as a whole. In present-day terminology, this would be described as a field of action of genes affecting not individual teeth, but the whole dentition, even though the effect of a given gene may be more apparent on one tooth than another.

A number of Bateson's contemporaries (e.g. Wortman, 1886; Scott, 1892) held opposing views, in which the units of the dentition were regarded as functional rather than morphogenetic; each tooth had become adapted to its particular function, independent of its neighbors. Supporters of the premolar analogy theory, for example, believed that premolars lag behind molars in evolution since, being further removed from the insertion of the muscles of mastication, they are less subject to selection pressures favoring the acquisition of the molar pattern. Indeed, Scott (1892) considered that the apparent serial homology between premolar and molar cusps was the deceptive result of convergent evolution, a view which implies that each tooth is under independent genetic control.

Rutimeyer (1863), remarking on the analogy between dental differentiation and that seen in other biological structures such as the leaves and floral parts of plants, suggested that all teeth were modified molars. Bolk (1922) translated this suggestion into ontogenetic terms by proposing that all teeth have the same potentiality for development but that this remains latent in simple teeth and fully realized in molariform teeth only. This proposal implies that all teeth develop in the same direction but to different extents, and that fully molariform teeth during their course of development pass through stages in which they resemble simpler teeth. A number of diverse observations lend support to this view: (a) The last cusp to develop in maxillary molars is commonly the distolingual cusp (hypocone), and this is the cusp most likely to be missing in reduced distal molars and premolars; (b) generally, the distal part of a molar develops after the mesial part and when the last molars are simplified, it is the distal part of the tooth that is principally reduced; (c) similarly, the first and second maxillary deciduous molars of *Homo sapiens* differ in that the former has a weaker development of the distal part of the crown than has the latter; (d) sometimes the first and second permanent molars are dwarfed and they then resemble premolars (Remane, 1926); (e) when premolars are reduced in *Homo sapiens,* it is the lingual cusp that disappears, the cusp that is absent from the anterior teeth.

But careful examination of the stages of tooth development indicate that Bolk's interpretation probably needs modification. For instance, studies of the development of the human deciduous dentition (Kraus and Jordon, 1965) have shown that the distal part of molars appears earlier and reaches a larger size than the more mesial parts of the tooth.

In most mammalian dentitions a gradation of tooth shape can be recognized in passing from one region of the dentition to the next. On the basis of this observation, Butler (1939) proposed, in his theory of tooth development, that there exists within the embryo a continuous morphogenetic gradient which determines the form of the tooth within it. More recently, Butler (1956) has reviewed the ontogeny of the molar pattern in relation to his *Field* theory; whereas, Henderson and Greene (1975) have examined this theory in relation to primate evolution. Lombardi (1975) has established that multivariate analysis is based on many tooth dimensions is required to examine critically the morphogenetic fields of the primate dentition. Only in this way is it possible to treat teeth as multidimensional units where the correlation among dimensions may be accounted for.

If such a continuous gradient exists, it follows that slight alterations in the relative position at which a tooth germ develops would lead to corresponding modifications in tooth morphology. This would be most obvious in regions where the gradient might be expected to change rapidly, such as that which lies at the boundary of the premolar and molar series. This expectation is borne out by the findings that the first maxillary deciduous molar, for example, shows varying degrees of molarization both in human ontogeny and among various extant and fossil hominoid forms. Such variation could arise from genetic modifications of the morphogenetic gradient in relation to the position of individual tooth germs or alternatively from genetically controlled changes in the way in which the tooth germs react to any given level of the field.

Dahlberg (1945) has applied the field concept to twelve specific features of the human dentition, namely general differentiations, size, division into regions, variability, general form, regional form, shovelling and the presence of five additional molar cusps. Of these, general differentiation, size and general form do not vary from region to region. For example, hereditary changes in the mineralized dental tissues, the overall size and the angularity of the teeth are characters equally expressed in all regions. The regional form of the tooth is strong and the variability is least in the first incisor, canine, first premolar, and first molar region. Shovelling is restricted to the incisor part of the field while the presence of the additional cusps also has a strong regional character: Carabelli's cusp, for example, occurring most frequently in the first molar while Bolk's paramolar cusp is usually found in the third molar.

Dahlberg also emphasizes that correlative or modifying influences come into action at varying times with varying effects. He suggests that the fields are likely to be at their peak when the first tooth of each group is differentiating. The influences might later be reduced so that distal members of the group are reduced or even absent. This suggestion which would explain the frequency of factors, as well as modifying the morphogenetic gradients which determine the fields and reaction of the tooth germs, could also act through changes in the correlative or modifying influences.

Most of the characteristics that have been investigated in

studies of dental inheritance belong to the category of quasi-continuous variables, i.e. they may be present or absent but when present they vary continuously from the lowest level of expression to the highest. The accepted model of quasi-continuous variation depends on the assumption that there is an underlying scale of continuous variation of some attribute produced by a combination of all the genetic and environmental factors involved. The character is present when the attribute exceeds a certain threshold level and absent when the attribute falls below this threshold. The more the attribute exceeds the threshold, the more intense the expression of the character is likely to be. In characters with this type of variation, subjective classifications of their degree of expression are necessary. If single autosomal genes are believed to be involved, for example, three classes of the type, "nonaffected," "minimally to moderately affected," and "moderately to maximally affected" may be recognized on the assumption that they correspond to the three genotypes produced by the segregation of two alleles, i.e. homozygous for one allele, heterozygous, and homozygous for the other allele. The agreement of the frequencies of these three hypothetical genotypes in a population with the expectation derived from the application of the Hardy-Weinberg law has then been taken as evidence that a single gene control does exist. In fact, such an approach is valid, as a means of identifying the nature of genetic control, only when applied to a known genetic locus, identified through a study of related individuals.

Many studies have been concerned with the inheritance of sexual dimorphism in tooth (usually crown) size. Porter (1972) detected a small sex difference in mean tooth dimensions derived from fifty-six separate measurements in a sample of Pima Indians. The spread of distributions was such that the upper extreme of the male distribution reduced by as much as a fourth of each distribution. Through a stepwise discriminant function analysis, a smaller set of twelve variables was selected from the original set of fifty-six, selection being based on the significance of contributions to the generalized distance (D^2) between the two sex subgroups. These twelve variables alone were enough to account for most of the distance between males and females. Univariate analysis of the data showed no significant sex difference in ten variables, of which three were among the twelve selected variables that contributed significantly to the sex differences in the multivariate analysis. These variables were the mesiodistal diameter of the mandibular right central incisor, left second premolar and the buccolingual diameter of the mandibular right lateral incisor. In turn, a majority of the variables (39) with highly significant sex differences did not contribute significantly to sex differentiation in multivariate analysis. This contrast in data arises from the correlation among tooth size diameter which, when removed, enables the male and female comparisons to be observed differently.

The twelve variables selected by the stepwise discriminant analysis are of interest. The mesiodistal diameter of the mandibular canine contributed the most among both the fifty-six initial variables and twelve selected variables in discriminating between males and females. The lateral incisors and second premolars made the next largest contributions. The right and left maxillary and mandibular homologous teeth seemed to contribute differentially to sex differences, and in no instance did both the mesiodistal and buccolingual crown diameters of the same tooth enter into this smaller set of significantly discriminating variables. If the shape of a tooth may be represented approximately by both the mesiodistal and buccolingual tooth diameters combined together, the results of this study appeared to suggest that, in the human dentition, shape does not display sexual dimorphism as much as size, since ten of these twelve variables were mesiodistal crown diameters.

In a recent study of the dental casts of 199 pairs of like-sexed twins, Biggerstaff (1975) measured the basal area of the molars. The results indicated a cusp-size hierarchy specific for molar type, i.e. five-cusped molars with a distal fovea and distal marginal ridge (5fd), five-cusped molars without a distal fovea and without a distal marginal ridge (5o), and four-cusped molars (4c). Sexual dimorphism was evident in cusp size in 5fd molar cusps but not in 5o molar cusps. However, males showed a significantly higher frequency of 5fd molars, whereas females had a higher frequency of smaller 5o and 4c molars, which had fewer crown components. Moreover, the 5o molars of females had cusps as large as or larger than 5o molar cusps in males. When observed, however, most intrapair differences for cusp size, using F ratios, indicated a low component of hereditary variability.

Sirianni and Swindler (1973) measured the mesiodistal diameters of the deciduous teeth on 135 related and unrelated pairs of *Macaca nemestrina*. To minimize bias, the left maxillary and mandibular teeth were measured and recorded without knowledge of one monkey's relationship to another. The average tooth size product-moment coefficient for full siblings was 0.38, whereas that for unrelated monkeys was close to zero, with minimum sexual dimorphism. This data, therefore, suggested that the X chromosome may not be involved in determination of tooth size, although the possibility exists, however, that the Y choromosome is involved.

The inheritance of tooth size in families has been reported by several workers, from examination of the mesiodistal and buccolingual crown diameters (Bowden and Goose, 1969). In general, significant correlations between parents and offspring and among siblings have been found. More recently, however, Lavelle (1972) reported low correlations in caucasoid, mongoloid and negroid families living in Birmingham. It was suggested that this finding may be due to the fact that in recent immigrant populations the parents were usually born abroad while the children were born in the new country. This clearly would increase environmental variation and reduce correlations. This suggestion gains support from the finding of lower correlations by Goose and Lee (1973) in a Chinese population in Liverpool than in a corresponding caucasoid group (Bowden and Goose, 1969).

Growth of Tooth Germs

Attempts to collect metrical data on tooth germ growth has involved measurement of serial sections (Gaunt, 1959), and direct measurement of extracted tooth germs (Kraus and Jordan, 1965).

Using the last method, Butler (1971) measured deciduous tooth germ growth in an ethnically and sexually heterogeneous sample of embryos and fetuses. An early period of rapid growth was described up to about the twentieth week, followed by a period of gradual decline in growth rate. Kraus and Jordan (1965) measured the mesiodistal length of deciduous incisor, canine and molar teeth in Amerindian and American Negro fetuses of known age and calculated the regressions of length against age for each tooth. They found a significant difference between the two races but this disappeared if the data for the period after twenty weeks is excluded and their values are then in close agreement with those of Butler for fetuses of mainly European descent. Combining the data of Kraus and Jordan with those of Butler provides information based on a total of about a hundred specimens for the growth of each deciduous tooth over an age range of twelve to twenty weeks The plots of tooth size against age are then very close to a straight line, with the deciduous molars, for instance, growing at a rate of between 0.4 and 0.5 mm per week.

The slowing of deciduous tooth growth at about twenty weeks has been attributed to the initiation of calcification. However, Butler (1971) noted that the mandibular first deciduous molar begins to calcify at fourteen to sixteen weeks, the second mandibular molar at eighteen to twenty weeks, and the mandibular first permanent molar at twenty-nine to thirty-two weeks, that is respectively well before, just before, and well after the phase of decelerating growth.

As the tooth develops and grows, it undergoes changes in proportion. When the heights of the cusps of the mandibular second deciduous molar are plotted against mesiodistal tooth length, for instance, a series of parallel lines is obtained for the major part of the growth period (Butler, 1968). Towards the end of the period, growth in length ceases before that in height so that the lines curve upwards. In both the mandibular deciduous molars and first permanent molars, the talonid begins to grow at a later stage than the trigonid, so that, at first, it is proportionately small but owing to its greater growth rate it eventually equals or exceeds the trigonid in length and width. By extrapolating the growth lines to zero, one can estimate the tooth length at which the various elements start growth. The order so obtained is similar to that of cusp appearance postulated by Cope-Osborn terminology.

During the early period of rapid growth, the intercuspal distances increase in proportion to tooth length but, at a later stage, the cusps tilt apart at a greater rate than can be accounted for by growth of the base on which they stand (Butler, 1967a). As the crown expands, the trigon and talonid basins occupy an increasing proportion of the area of the crown. This process is presumably due to growth in the uncalcified areas between the cusps and must consequently end when the calcified areas of the cusps unite. Hence, the amount of spreading of the cusps will depend upon the incidence of calcification so that the order of union of the cusps plays a part in determining the final tooth form.

Before calcification of a cusp begins, its tip undergoes a characteristic sequence of changes (Butler, 1967a). Starting as a flattened area, it first becomes rounded then increases in height and develops a point which becomes progressively more acute as development proceeds. A similar process takes place along ridges which also become sharpened before calcification. With tooth calcification, the contours tend to become more rounded again.

QUALITATIVE ASSESSMENT OF TOOTH SHAPE

The many similarities in their dental morphology provide some of the strongest reasons for grouping hominids and pongids in a single superfamily, but the greater proportion of the very large literature dealing with this topic has dwelt on the contrasts in their dentitions. The majority of studies, especially those dealing with nonhuman primates, are entirely descriptive in nature. This has led to a considerable divergence in the conclusions reached by workers who have employed dental features as aids in assessing phylogenetic lineages within fossil members of the Hominoidea. An examination of some of the qualitative comparisons that have been made will illustrate why such difficulties persist.

Incisor Form

The incisors in almost every known primate have been reduced in number from the primitive eutherian condition of three in each quadrant. Indeed, the possession of two incisors was taken by Linnaeus as a diagnostic feature of the *Order* but several exceptions occur among the prosimians. The tree shrews, for example, retain three mandibular incisors while the tarsiers have usually only one incisor in the lower jaw. In recent genera of lemurs, the mandibular incisors are specialized to form a comb-like structure and the upper incisors have dwindled to mere rudiments. Within the Anthropoidea, however, there is constancy of number and form in the incisor series. As a result of this constancy, the incisors have played a rather minor role in attempts to determine hominoid affiliations.

Perhaps the most frequently used feature of the incisor teeth in studies of human populations has been the presence or absence of incisor shovelling. In this condition, the incisor has prominent mesial and distal ridges on its lingual surface enclosing a central fossa. It is usually bilateral although it occasionally occurs unilaterally (Hrdlicka, 1920). In order to add precision to such studies, attempts have been made to quantify the degree of incisor shovelling: (Hrdlicka, 1920), for example, used four categories, namely shovel, semishovel, trace-shovel and no-shovel (see Table II).

Moderate degrees of shovelling have been noted to occur more frequently in earlier compared with later populations of *Homo sapiens* (Carr, 1960) and a certain amount of shovelling has been described in the teeth of Neanderthal man (Vallois, 1950), *Homo erectus* (Weidenreich, 1937), and the australopithecines (Robinson, 1956). The origin of the trait is obscure. Nevertheless, from examination of parents and sibs, and sibs-sibs for incisor shovelling, Blanco and Chakraborty (1976) have computed that for the transmission of shovel shape, 68 percent of total variability may be ascribed to the additive effect of genes.

Other methods of comparing incisor crown form include the classification of Williams (1928) into square, tapering and ovoid, and the method of Goose (1956) based on the description of the labial surface of the tooth as an approximation to a trapezium. A measure of quantification is introduced in the latter method by the calculation of three attributes of the trapezium:

(a) The surface areas in mm $^2 = \frac{(\text{height} \times \text{mesiodistal and cervical widths})}{2}$

(b) the tooth width index $= \dfrac{(\text{mesiodistal width} \times 100)}{\text{crown height}}$

(c) the angle of taper $= 20(\tan = \dfrac{\text{mesiodistal} - \text{cervical width}}{2 \times \text{height}})$

Neither of these methods has yet received wide application and both, of course, remain ultimately subjective in their nature.

Canine Form

Canines are much more variable than incisors at all levels of primate evolution. In many groups, the teeth are hypertrophied often with pronounced sexual dimorphism as, for example, in the maxillary canines of lemurs and both maxillary and mandibular canines of larger monkeys and apes. In other groups, the canines have been lost, as in *Daubentonia*, or assimilated into the incisor series. The development of an incisiform appearance has occurred in the specialized mandibular dentition of modern lemurs and, to a more limited extent, in both maxillary and mandibular dentitions of the genus *Homo* where canines range between spatulate and pointed forms.

One of the traditional assumptions of evolutionary studies of the human dentition is that during the emergence of man, canine size has been progressively and drastically reduced. Remane (1927) listed a number of characteristics that are generally associated with large crowns in this tooth and yet are found in the small tooth of modern man. Perhaps the most widely quoted (see, for example, Clark, 1967) is the length of the root of the maxillary canine which is much longer than would be expected from the assumed anthropoid characteristic of positive allometry between root and crown size. In fact, the relationship between the sizes of the root and crown seems to be quite variable in the primates generally. To take just two examples, the generalized Eocene form Omomyidae had canines with small crowns but roots longer than those of adjacent teeth and the roots of the low-crowned mandibular canines of *Propliopithecus* were also relatively long. The relationship between crown and root size in canine teeth may, therefore, be too complex to allow simple and direct deductions to be made about crown size, present or past, from root size.

The tendency of the human upper canine to erupt after the premolars (Schultz, 1935) is another observation which is used as evidence for a formerly larger size of this tooth since it mirrors a similar tendency seen in catarrhine monkeys with large canines (Schultz, 1956). But the data of Dahlberg (1948), indicates that delay in canine eruption may be merely part of the normal range of variation in the timing of eruption in modern man.

The canines in both Neanderthal man and *Homo erectus* are larger, on average, than those of contemporary mankind. The argument for hominid canine reduction based on these observations would lose much of its force if *Australopithecus*, with a canine probably no longer than that of modern man (Leakey, 1962), should eventually prove to be an ancestral hominid. The rather larger canine in some Pleistocene forms would then prove to be no more than a relatively minor variation, probably part of the enlargement of all the anterior teeth observed during the Middle Pleistocene (Brace, 1976b) and the slight decrease in canine size which has occurred in *Homo* since this time, an accompaniment of the reduction in the size of the molars and premolars (Bilsborough, 1969).

Most Primates exhibit strong sexual dimorphism in canine size (Schultz, 1969). Dimorphism is well developed in terrestrial Old World baboons (Swindler, McCoy and Hornbeck, 1967) as well as in the arboreal New World howler monkey, *Alouatta* (Zingeser, 1967). Even in man and gibbon in which sexual dimorphism in canine size is small, there is a significant difference of 3 to 10 percent in man (Moorrees, 1957) and 6 to 12 percent in gibbon (Frisch, 1963). Kinzey (1972) reports the complete lack of sexual dimorphism in the canines of *Callicebus moloch*. The large canines of males of other nonhuman primates apparently serve as defense and display and/or intragroup aggression organs, whereas *Callicebus moloch* exhibits an extremely low level of intraspecific aggression (Mason, 1966). The positive selective value of small canines is not, however, understood. It may serve to permit a more efficient grinding mechanism through a greater range of side-to-side jaw movements (Mills, 1963) and to provide a functionally more efficient horizontally sharp shearing device

(Every, 1970), since the tip of the canine wears flat in *Callicebus*, not to a point as in the canine-honing mechanism of other non-human primates (Zingeser, 1969). Similar reasoning has been put forward in the parallel development of small canines in human evolution.

Cheek Teeth

The premolars of primates show a tendency towards number reduction. Of the original four teeth in each quadrant, P1 was eliminated very early in primate evolution. In Indri and in all Old World members of the Anthropoidea, P2 is also missing. Associated with this numerical reduction, the remaining premolars, especially P4, have become enlarged and complicated by the upgrowth of cusps from the basal cingulum so that they assume a somewhat molariform configuration.

Generally, the Anthropoidea show only moderate molarization of the premolar teeth. In lemurs, the process is more pronounced so that the demarcation between molars and premolars is less than in the Anthropoidea. The maxillary premolars of the anthropoid apes are bicuspid with the lateral cusp being the larger. In the mandibular dentition, the posterior premolar is bicuspid with approximately equal sized cusps and a well-developed talonid while the anterior tooth is unicuspid. The distinctive sectorial form of this latter tooth appears to have been already present in Miocene and Pliocene pongids. Within the Hominidae, the anterior mandibular premolar is bicuspid, the cusps being of approximately equal size in *Australopithecus* and *Homo erectus* while, in modern man, the lingual cusp is usually reduced. Molarization of the premolars is especially marked in *Australopithecus robustus* and *Giganthopithecus*.

The number of molar teeth has remained constant, at three in each quadrant, in all Primate genera except the marmosets in whom the third molars have been lost. The form of these teeth also shows considerable constancy although some species of Ceboidea as well as *Homo sapiens* display a tendency towards reduction, or even disappearance of the last of the series. The configuration of the cusps and grooves of the permanent mandibular molars bears many similarities in man and the anthropoid apes. The fundamental configuration (the dryopithecus pattern) has five main cusps and a Y-shaped groove. As its name suggests it is found in the molar teeth of all known species of the extinct ape, *Dryopithecus* (Gregory, 1922).

Gregory (1922) in defining the ancestral mandibular molar pattern of man wrote, "The pattern of lower molars of *Dryopithecus* may be broadly described as follows: There are five main cusps, three of which (protonconid, hypoconid, mesoconid (hypoconulid) are on the external side of the crown and two (metaconid and eutoconid) on the internal side. The metaconid which is the higher cusp, is directly internal to the protoconid; the hypoconid is opposite the valley between the metaconid and entoconid; the mesoconid (or hypoconulid) is on or near the posteromedian border of the tooth behind the hypoconid and eutoconid . . . The surface of the lower molar crown is likewise characterized by the arrangement of certain furrows (grooves) which converge into a prominent inverted V ($\wedge$), the narrow end of which is at the center of the crown. From the furrow ends and sides of this truncated form, other furrows radiate as follows: (a) an anterior central furrow between the protoconid and the metaconid, (b) a posterior central furrow between the mesoconid and the eutoconid, and (c) one or two internal furrows between the metaconid and the eutoconid." Gregory asserted that comparison of mandibular molars in both the dryopithecine and human deciduous and permanent dentitions provided clear evidence that man was more nearly allied to the Mid-Tertiary *Dryopithecus* and more remotely to the existing chimpanzee and gorilla.

During hominid evolution, this presumably basic molar cusp pattern has undergone a variable amount of change, any departure from the Y5 configuration being generally regarded as a later stage in molar evolution (Korenhof, 1960). In one variant (+4), the crown bears four main cusps separated by a cruciform groove, the hypoconulid having been eliminated (Schuman and Brace, 1954). A further variation is the X pattern (Jorgensen, 1955), in which the four cusps meet approximately at a point at which the pairs of grooves surrounding the protoconid and entoconid each form an acute angle. Other variations encountered are the Y4 and +5 patterns.

The frequency of these patterns differs between human populations and also between the three mandibular molar teeth. In a sample of North American caucasoids, for example, 87 percent of individuals had a Y5 pattern in M1, while in M3, 62 percent had a +4 arrangement (Hellman, 1928). The corresponding figures for a Danish sample (Jorgensen, 1955) were 67 and 34 percent. In a British caucasoid sample, Lavelle, Ashton and Flinn (1970) found the percentage frequencies of Y5, Y4, +5 and +4 were 72, 4, 23 and 1 respectively for M1, 2, 4, 6 and 88 for M2; and 4, 2, 34 and 60 for M3.

A large number of recent pongids also exhibit a + groove pattern. In the chimpanzee for example, there is a + groove incidence of 22 percent in M, 76 percent in M_2 and 100 percent in M_3 (Schuman and Brace, 1954). Among fossil hominids, *Australopithecus* (Robinson, 1956) and *Homo erectus* (Weidenreich, 1937) have incipient + groove patterns while Neanderthal skulls often show a Y configuration on M1 and a + configuration on M2 and M3 (Weidenreich, 1937).

Since the form of the mature crown represents an expression of the sum of the ontogenetic processes, cusp size and cusp number are probably the two most significant polymorphisms. Crown size is dependent upon the growth patterns of the individual cusps and ridges. The groove configuration is just a visual manifestation of the terminal boundaries of a mature cusp or ridge being determined by the size and position of the individual crown components. Hence, the groove configuration is not a direct manifestation of the genetic complex, but a record of the inability of the ameloblasts to continue the deposition of enamel within the limited space between two or more contiguous crown components.

Biggerstaff (1968), therefore, suggested abandoning the Y and + categories, and substituted instead the following categories for the lower molars: (a) five-cusped molars with a distal marginal ridge and a distal fovea, (b) five-cusped molars without a distal marginal ridge and a distal fovea, and (c) four-cusped molars. These categories have the virtue of easy definition but have yet to receive universal acceptance.

Morris (1970) has questioned the use of the dryopithecus pattern and its variations in studies of dental evolution on the grounds that the term Y5 and its variations lack specificity and fail to take account of the occlusal surface polymorphism known as the "deflecting wrinkle" (Weidenreich, 1945). This is an occlusal ridge located on the metaconid and extends from this cusp distalward to the central occlusal pit. He points out that in Jorgensen's (1955) illustrations of the Y and + patterns, the major difference between them is the presence of a deflecting wrinkle in the case of the Y pattern and the absence of the feature in the + pattern. In fact, however, as Morris illustrates, the Y pattern may exist in the presence of a deflecting wrinkle.

Many workers have enumerated the main cusps on the permanent mandibular molars in various human ethnic groups. Even with such an apparently clear-cut feature as cusp number, however, difficulties have arisen largely because of the subjective element in deciding what actually constitutes a main cusp. Nonetheless, certain broad conclusions can be drawn. In man, the mandibular molars tend to have four to five cusps in contrast to the five or six of ape teeth. Furthermore, while there is no significant variation in cusp number between the dentitions of the extant apes, the frequency of human molars bearing four major cusps is significantly greater in caucasoids than in mongoloids, negroids, or australoids (Lavelle, 1971).

The relationship between tooth size and number of cusps or groove pattern is controversial. One view is that the cusps and grooves vary independently of the crown dimensions (Jorgensen, 1955), but there is evidence that four-cusped mandibular molars are smaller than those with five cusps (Lavelle, Ashton and Flinn, 1970). Further adding to the complexity of the situation is the observation that the groove configuration is independent of cusp number. It is difficult, therefore, to believe that any simple relationship exists between the factors, as still unknown, which control these three attributes of crown morphology. Although crown size may represent an adaptation to the diet, Bailit and Sung (1968) have stressed the lack of any convincing selective value in either cusp number or groove pattern of human molars. Also there appear to be marked deviations from the anthropoid progression of molar size in pongids (Mahler, 1974).

Carabelli Trait

The Carabelli trait (Carabelli, 1842) is described as a cusp, cuspule, tubercle or other anomaly that is often observed on the lingual surface of the protocone of human maxillary permanent molars and maxillary deciduous second molars. Meredith and Hixon (1954) provide further synonymous terms, such as elevation of enamel, fifth-lobe, supplemental cusp, accessory cusp, mesiolingual elevation or prominence. The terms *Carabelli complex or Carabelli trait* have been used by Robinson (1956) and Kraus (1959) to indicate both positive and negative, e.g. pit or groove features that may occur on the mesiolingual aspect of the protocone. The Carabelli trait may be found on all three molars of the maxillary permanent dentition. Usually, although not always, it is present on both sides and may be symmetrical or asymmetrical in development.

There is still controversy about whether the Carabelli trait developed *de novo* or whether it arose from the lingual cingulum. Adloff (1908) took the latter view in which he was supported by Robinson (1956). De Terra (1905), on the other hand, considered the Carabelli trait to be an outgrowth of the crown separate from the cingulum. A distinction between the origin of a pit and a tuberculum was made by Weidenreich (1937). He believed that the pit represents "a remnant of the cingulum common to all primates, including the anthropoids, and the tubercle is an accidental variation without phylogenetic significance."

The original form of the lingual cingulum may be represented by the condition seen in *Adapis* (a lemuroid prosimian from the Middle and Upper Eocene). Here the cingulum is present along the lingual and mesial slopes of the protocone. This structure is more extensive in later lemurs and in both catarrhines and platyrrhines (Korenhof, 1960) but in the dryopithecine fossils it appears to have undergrone a great reduction (Gregory and Hellman, 1939), especially in the Eastern forms (Frisch, 1965).

In extant pongids, the lingual cingulum is best developed in

Gorilla and least in *Pongo* (Schuman and Brace, 1954). Whether or not a Carabelli trait occurs in these groups is controversial. De Terra (1905) stated that it did not but Campbell (1925) found a small groove, which he took as an indication of the trait, in three out of twenty-one gorilla molars. Weidenreich (1937) also reported the presence of pits on the lingual surface of maxillary molars in this animal. Carabelli traits in pongid teeth have also been reported by Ashton and Zuckerman (1950). The balance of opinion appears to be, therefore, that features which can be termed Carabelli traits do occur in living apes but are probably less pronounced and less frequent than in modern man.

Traces of the lingual cingulum are still evident in fossil hominids. Robinson (1956) has described such a cingulum in australopithecines where it is more marked on the first compared with the third molar. A small Carabelli trait has been noted on the maxillary third molar of *Homo erectus* by Korenhof (1960), although this was not described in the Chinese fossils of this species by Weidenreich (1937). A Carabelli trait or cingulum was not found in Neanderthal specimens by Bay (1958) but McGown and Keith (1939) have described negative structures which are possibly related to the Carabelli trait in some fossils of this group. Clearly much of this confusion is due to inconsistency in terminology which derives, in turn, from uncertainty about the origin of the Carabelli trait and its relationship to the lingual cingulum. Until this has been resolved, it seems likely that disputes about its true frequency in any group will continue.

Although the Carabelli trait is usually assumed to be of importance in taxonomic studies, (Rajcinova, 1972) its functional significance is obscure. De Terra (1905) has argued that the presence of the Carabelli trait on maxillary first molars tends to enlarge the occlusal surface, compensating to some extent for the loss of tooth structure associated with reduction of the second and third molars. Dahlberg (1963) has suggested that such compensatory addition of tooth material may have been of adaptive significance, in natural selection. However, Moorrees (1957) has argued that the combination of frequent absence of third molars and low incidence of Carabelli's trait seen in mongoloids generally fails to support this view. De Jonge (1958) suggested that reduction and elimination of Carabelli's trait is related to the simplification of the occlusal surface as exemplified by hypocone reduction. Keen (1968) examined the incidence of the Carabelli trait in five categories of manifestation (smooth lingual surface, variable pits, grooves or depression, slightly developed cusps, moderately developed cusp, and large cusp) in nearly 400 male caucasoids in relation to cusp number, mesiodistal diameters of the maxillary first and second molars and the presence or absence of third molars. The principal conclusions were that: (a) the total frequency and degree of expression of Carabelli trait on the first maxillary molar were reduced when there was unilateral or bilateral third maxillary molar agenesis; (b) the maxillary second molar with three cusps exhibited fewer and less pronounced expression of the Carabelli trait than those with four cusps; (c) first maxillary molars with three cusps were rare and showed no evidence of the Carabelli trait; (d) the total frequency and degree of expression of the Carabelli trait on the first maxillary molar were reduced when the adjacent second molar was tricuspal; (e) the mean mesiodistal crown diameter of the maxillary first molar with no Carabelli trait was smaller than those demonstrating the trait; (f) the maxillary second molars which were adjacent to the maxillary first molar showing no Carabelli trait were smaller in mesiodistal crown diameter than those which were adjacent to the maxillary first molar expressing the trait. Keene's study would therefore appear to indicate that the Carabelli trait and hypocone are similar in their variability and the expression of each is in some degree dependent upon the factors which are involved in the structural and numerical reduction observed in the maxillary molars.

There is no consensus of opinion regarding the heritability of the Carabelli trait, although the study of its genetic basis has received more attention than that of any other dental trait (Lasker, 1951). Twin studies provide ideal data for genetic studies in that the variability between monozygotic and dizygotic twin pairs gives an estimate of the degree of hereditary variation and takes account of the positive correlation between the genetic relationships and the similarity of the environment. Using such twin studies, Biggerstaff (1973) has shown that the Carabelli trait does not have a high degree of heritability, which implies that previous studies using this feature in taxonomy must be treated with caution.

Paramolar Tubercles and Protostylids

The paramolar cusp comprizes any stylar cusp, supernumerary inclusion, or eminence on the buccal surface of upper and lower premolars and molars (Dahlberg, 1945), and has been described in most modern human populations (De Jonge-Cohen, 1958). The protostylid is derived from the anterior part of the buccal cingulum of mandibular molars. Varying degrees of prominence have been noted in *Australopithecus* (Dart, 1948), *Homo erectus* (Weidenreich, 1937), Pima Indians (Dahlberg, 1945), Japanese (Suzuki and Sakai, 1954) and in occasional isolated individuals of many modern human populations (De Jonge-Cohen, 1958), appearing to be associated with retention of a primitive molar pattern, i.e. retention of the dryopithecine configuration). For both the paramolar tubercles and protostylids, however, the subjectivity in deciding whether such features are present or absent limits their value in making taxonomic judgments.

Taurodontism

This is a condition in which the crown is enlarged relative to the length of its roots and is frequently associated with enlarged pulp chambers. The condition has ben described as a distinct trait in Aleuts (Moorrees, 1957) and Eskimos (Pedersen, 1949) as well as in Neanderthal man and other fossil types (Howell, 1963). Blumberg, Hylander, and Goop (1971), however, describe taurodontism as a continuous variable controlled by polygenes without discrete modes of expression. Also, enlarged pulp chambers, which may or may not be associated with taurodontism, have been described as predominating in early man, with a gradual reduction through Pleistocene (Howell, 1967). Neither the functional nor evolutionary significance of taurodontism has been elucidated, and this coupled with the difficulty in diagnosis, especially in the absence of radiographs, precludes the value of this feature in evolutionary studies.

ODONTOMETRY AND QUANTITATIVE TOOTH DESCRIPTIONS

Crown Measuration

Several factors limit the accuracy of tooth measurements. These include the complexity of tooth shape, attrition, and the definition of datum points. To these must be added the general confusion that exists in the literature regarding the dimensional nomenclature.

The mesiodistal diameter, which is variously referred to as tooth "breadth," "length," or "width," is most frequently defined as the distance between the contact areas, although the mesiodistal diameter at the cervical margin is more constant since it is not affected by attrition. There is more general agreement that the best definition of the buccolingual (or labiolingual) crown diameter is the greatest diameter of the tooth in the buccolingual plane, at right angles to the mesiodistal crown diameter. Nevertheless, in the case of molars, there is no general agreement whether this dimension should be recorded for the trigon or talon. Furthermore, the measurement of cusp heights or intercusp distances do not seem to figure much in odontometry, perhaps since for such measurements the datum points are so variable due to attrition.

Attrition is, of course, not limited to the occlusal surface and may lead to inconsistencies in the measurement of any crown dimension. Although it is obviously preferable to measure only teeth with no attrition, a number of arbitrary methods have been devised to compensate for the loss of tooth substance due to attrition, but such methods have yet to withstand critical analysis.

Various attempts have been made to estimate overall crown shape by using indices based on mesiodistal and buccolingual diameters: crown module, average of the mesiodistal and buccolingual diameters, crown index, ratio of the mesiodistal and buccolingual diameters, robustness, product of the mesiodistal and buccolingual diameters. All such indices are rather crude approximations of such a complex shape as a tooth, so that teeth with different shapes may possibly share a similar index.

Whether using indices or individual crown dimensions, problems may arise from correlations existing between the dimensions of the maxillary and mandibular teeth, and between teeth from the right and left sides. The degree of these correlations varies between different human populations (Gabriel, 1955; Bolton, 1958), so where possible, it is preferable to compare dimensions derived from homologous quadrants and to make due statistical allowance for correlation when this is appropriate.

As already stated, the use of mesiodistal and buccolingual diameters alone provide only approximate indications of crown shape. A more complete definition would require the use of a large number of datum points. Such points have been defined in *Homo sapiens* for the mandibular premolars (Kraus and Furr, 1953; Ludwig, 1957), molars (Lavelle, 1971) and premolars and molars (Biggerstaff, 1969). Wakatsuki (1967) has used the shape of contour lines on the crown surface at intervals of 2 mm to contrast the crown shape of the deciduous dentition, whereas Hine, Lavelle and Flinn (1970) used three-dimensional coordinates of many homologous datum points to compare the mandibular first and second deciduous teeth in man and apes. More recently, Clark, Lavelle, and Flinn (1971) have employed the technique of stereophotogrammetry to record the three-dimensional coordinates of 400 datum points per tooth to a high degree of accuracy in the comparison of the mandibular first permanent molars of *Homo Sapiens* and *Pongo*. Such sophisticated methods undoubtedly provide ways of defining accurately complex shapes, but their use in comparative studies either within or between primates has so far been limited.

Tooth Roots

Quantitative assessment of tooth root morphology has received little attention compared with crown form. One of the principal deterrents being the lack of clearly defined datum points. Nevertheless, Kovacs (1971) has computed the ratio of the surface areas of the crowns and roots of teeth; the value of the mandibular canine was 154.2 percent in man, 217.2 percent in *Pan*, and 285.7 percent in *Gorilla*, the respective values for the mandibular first molars being 128.3, 158.3, and 166.4 percent. Nevertheless, until further datum points have been defined, it is difficult to ascertain how further quantitative data can emerge from the study of tooth roots see Chase and Swindler (1974) and Nicholls, Daly and Kydd (1974).

Dental Variability

Primates are a particularly variable mammalian Order, although there are conflicting views about the degree of variability in different species. The most obvious sources of intrageneric morphological variation lie in age changes and sexual dimorphism. In addition, a considerable degree of individual variation occurs within any age or sex subgroup. In serial structures such as teeth, there is also the possibility that variation may differ between one member of the series and another, e.g. in *Homo sapiens*, the degree of dental variation increases distally (Garn and Lewis, 1962).

The hypothesis of a strong genetic component in the development of crown size has long been suggested (Korkhaus, 1930), and tooth size is generally regarded to be independent of environmental factors (Osborne, 1967). The exact proportions of the genetic and environmental components of tooth size, however, remain unknown, although there is undoubtedly at least a measure of environmental influence upon the development of crown size (Goose, 1967).

One of the main difficulties in investigating this problem is the lack of a genetic model for crown size. Tooth size is subject in part to X-linked inheritance (Lewis and Grainger, 1967). There is definite evidence for communality in the relationship of corresponding tooth size in brothers and sisters (Garn, Lewis and Kerewsky, 1967), and for the relationship between sexual dimorphism in tooth size and body size (Garn, Lewis and Kerewsky, 1965). Attempts have been made to determine whether there is a general size factor linking genetic control of mesiodistal and buccolingual crown diameters, and Garn, Lewis and Kerewsky (1968) have found that posterior teeth are more significantly correlated than anterior teeth.

A complicating factor in investigating the genetic determination of tooth size is that of environmental interaction. Permanent teeth could well be influenced by diet and nutrition (Hunt, 1960), although there are closely related subgroups of *Macaca*

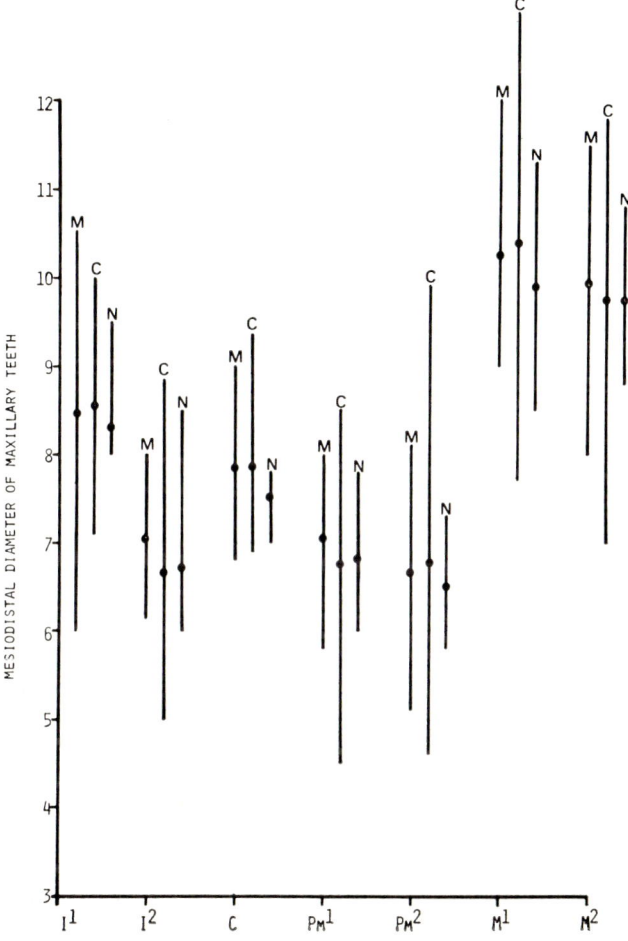

Figure 14. Variation in mesiodistal crown diameters of maxillary teeth in three ethnic samples: M = mongoloid; N = negroid; C = caucasoid: means and twice their standard deviations.

with significant differences in tooth size yet apparently sharing identical diets (Swindler, Gavan and Turner, 1963).

Correlations among tooth sizes are of interest, since correlation coefficients measure the degree of association between variables and indicate how effectively one variable can be predicted from another. Most dental correlations have been based on permanent teeth and their deciduous predecessors. Generally, ac-

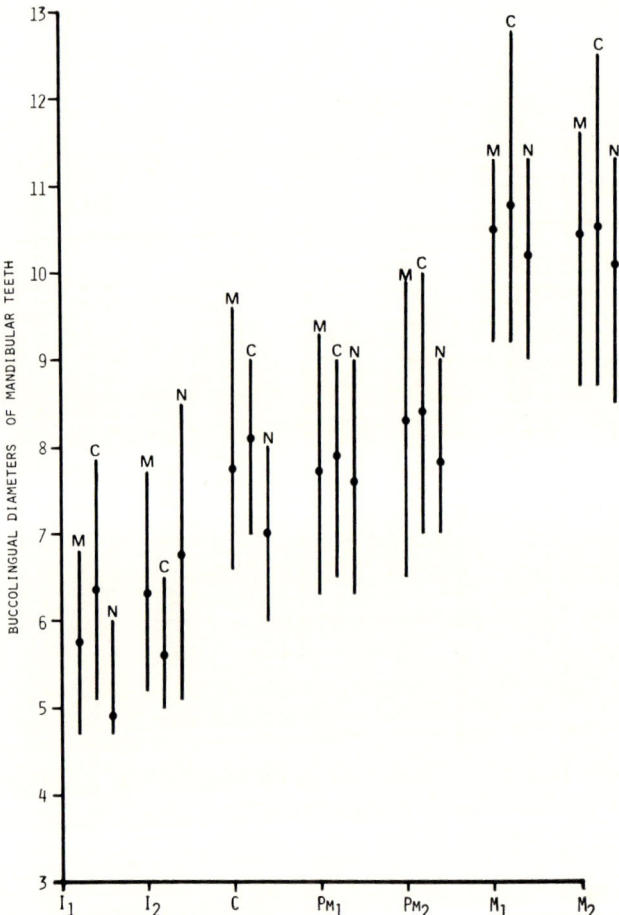

Figure 15. Variation in buccolingual crown diameters of mandibular teeth in three ethnic samples: M = mongoloid; N = negroid; C = caucasoid: means and twice their standard deviations.

curacy of predictions based on size of individual teeth or even groups of teeth, is quite low, as accuracy of prediction depends upon the square of the correlation coefficient. Thus, a correlation coefficient of 0.70+ is required to account for even 50 percent of the variability, and most tooth size correlations are well below that value. Prediction is, however, not the only use for correlations. Correlations between tooth sizes may be taken as measures

The Primate Dentition 155

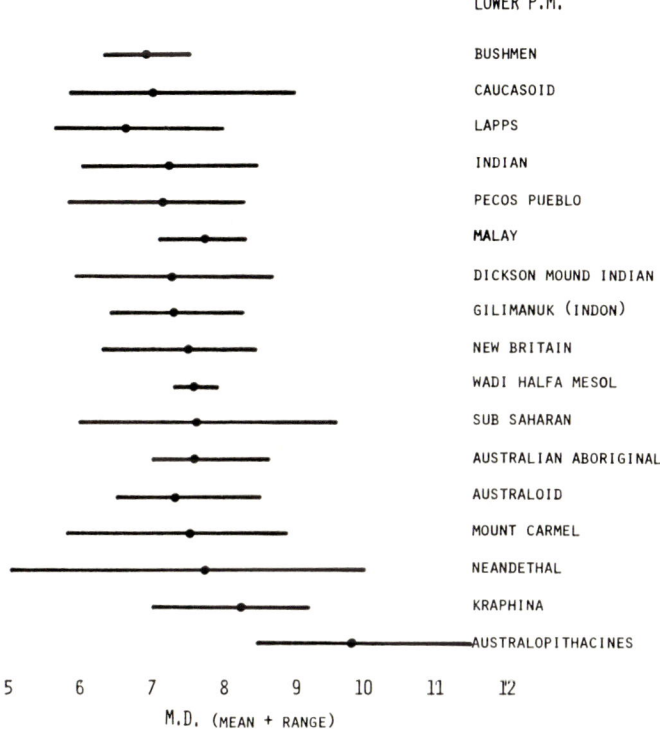

Figure 16. Variation in mesiodistal crown diameters of hominoids for the lower first premolar teeth: means and ranges.

of common genetic and environmental factors affecting tooth size. These common factors may vary between pairs of teeth and also may vary between different populations. Some investigators (e.g. Seipel, 1946) have examined interrelationships among various groups of teeth. Nevertheless, the only conclusion to emerge is that while correlations among tooth sizes may not differ completely with different populations, they differ enough that correlations derived from one population cannot be taken as representative of other populations. For the most part, correlations have been computed only between isolated teeth, whereas it might be far more informative if correlations were computed for the dentition as a whole. Thus, although correlation coefficients can be of undoubted value, much greater exploitation is required,

156 *Evolutionary Changes to the Primate Skull and Dentition*

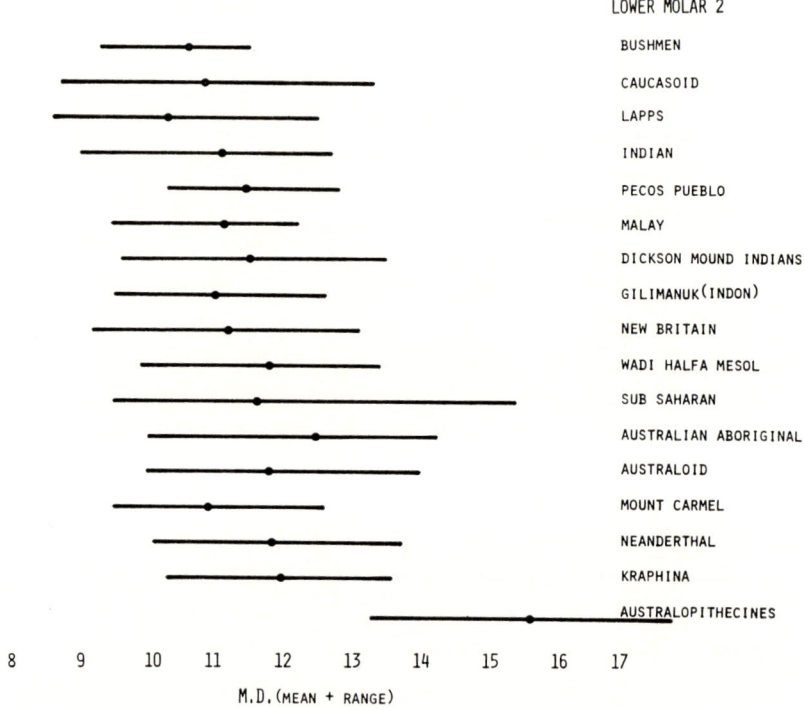

Figure 17. Variation in mesiodistal crown diameters of hominoids for the lower second permanent molar teeth: means and ranges.

particularly in order to ascertain whether the degree of correlation differs between various primate genera.

Various methods, additional to those used in standard univariate statistical procedures, have been devised to compensate for variability between dimensions. The coefficient of variation (standard deviation/mean) has been used extensively since it is presumed to allow comparison between populations with different means. There is, however, no a priori reason why, in such instances, the coefficients should give comparable values, so that their use should possibly be confined to comparisons of the same dimensions in closely related species. An alternative method is to examine the variance of the logarithms of sample means, although this transformation underestimates the difference in variability when the coefficient of variation is low and overestimates the dif-

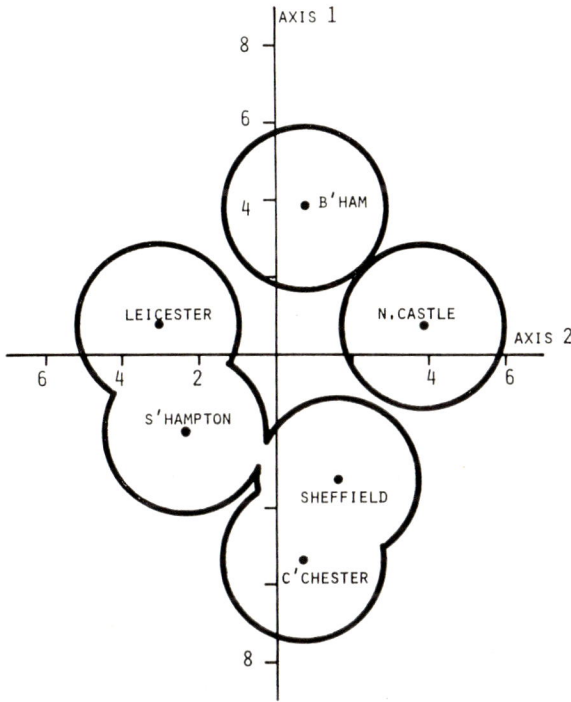

Figure 18. Variation of tooth size for British caucasoids from different regions of England. First two axes of the canonical analysis of the mesiodistal and buccolingual maxillary permanent tooth dimensions combined together: centroids and 90 percent confidence limits.

ference when the coefficient is high. A further method is to divide the range (maximum to minimum values) by N × standard deviation, so that the relative variances of different samples can then be estimated by comparing the actual with the expected range.

With the development of multivariate techniques, the statistical treatment of tooth dimensions has belatedly undergone a number of significant advances. For example, Carr (1960) used the *Coefficient of Racial Likeness* to distinguish between the dentitions of Middle Minoans with those from other ethnic groups, and discriminant functions have been used to compare the dentitions of human populations. Possibly the most notable example is that of Ashton, Healy and Lipton (1957) who showed

158 *Evolutionary Changes to the Primate Skull and Dentition*

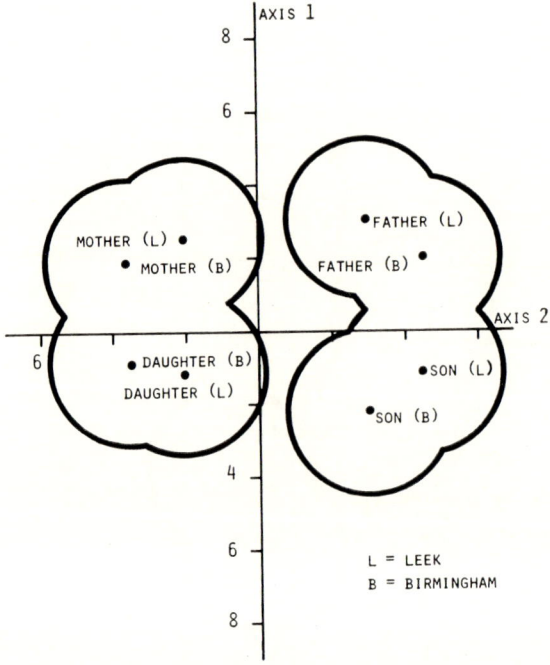

Figure 19. Variation in secular changes of tooth size in Birmingham and Leek, two British caucasoid population samples. First two axes of the canonical analysis of the mesiodistal and buccolingual maxillary permanent tooth dimensions combined together: centroids and 90 percent confidence limits.

that canonical analysis separated the canines and mandibular first premolars of man and apes, although there was less discrimination between the incisors and even less discrimination between the second premolars and molars. Using an identical technique, Lavelle (1971) reported that the mandibular molar dimensions are similar in mongoloids, negroids, australoids, and Anglo-Saxon caucasoids, but that there was considerable discrimination between these four groups and present-day caucasoids with sixteenth to seventeenth century caucasoids being intermediate.

However sophisticated the statistical analysis, the choice of dimensions to be measured is still based upon a subjective assessment of parameters which will emphasize the best differences or similarities in shape. The alternative is to measure very large numbers of dimensions which, until it is automated, is both time consuming and tedious. Thus, although teeth are the primary structures used in primate taxonomy, reliance on such structures must await the development of more critical methods.

Potter and Nance (1976) have recently partitioned bilateral tooth measurements in twins into three orthogonal contrasts, each associated with one degree of freedom, to estimate three parameters: discordance, asymmetry and mirror imagery. The results provided strong evidence for the existence of significant genetic determinants of almost all of the individual tooth dimensions, but little or no evidence for a genetic basis of asymmetry. The analysis gave no indication that monozygotic twinning was associated with an increased degree of either fluctuating asymmetry or mirror imagery when compared to dizygotic twins. Also, the data on monozygotic twins suggested that for most variables examined, the increment of environmental discordance resulting from the twinning phenomena was greater than the developmental noise that caused asymmetry within individual co-twins. Furthermore, the correlation among tooth dimensions appeared to be primarily of genetic origin, probably attributable to the pleiotropic action of either independent genes or groups of genes. Among the genetic factors that were identified, one appeared to affect the maxillary teeth in general, while a second influenced primarily the anterior mandibular teeth. The genetic determination of the maxillary and mandibular dentition seemed to be independent of each other, and a wider range of genetic factors were found to influence the mandibular than maxillary teeth. This suggested that a differential degree of evolutionary stability may have been achieved in the teeth of the two jaws. Nevertheless, further work on these lines is required in order to partition the genetic component of tooth variability.

EVOLUTION OF HOMINID TOOTH FORM

Within a species, teeth tend to vary allometrically one with another. With variation in absolute body size, the lever arm of the jaws tends to vary in direct relation with linear body measurements, while the occlusal surface of the teeth and muscle power will vary in relation to the squared values, and volumes of ingested food in relation to the cubed values of body measurements. Adaptation thus requires complex changes between different parts of the dentition as body size changes. It is clearly advantageous for the appropriate allometries to be built into the genotype, especially in forms that show extensive evolutionary plasticity. Allometric relationships between teeth would imply that they are positively correlated with each other with regard to size. There are many phyletic instances in which it would seem that some teeth increase in size at the expense of their immediate neighbors. Generally, however, teeth are positively correlated in size, the correlation tending to be greater between closely neighboring teeth than those more distantly placed, and maximally between teeth which actually occlude with each other (Van Valen, 1963). It is evident that a positive correlation will inhibit changes of a compensatory type, although compensation may occur if the allometric axis becomes transposed through a genetic change affecting the growth field.

Over the years, there has been considerable controversy about the evolutionary trends in hominid tooth form (see, for example, Keith (1925) Robinson, (1956) and Brace (1962) for some of the more critical studies from the extensive literature dealing with this topic). In 1971, Wolpoff presented a study of the phylogenetic trends in the hominid dentition which provides the most extensively documented investigation yet published. It is based on the mesiodistal and buccolingual dimensions (either taken by Wolpoff or extracted from the literature) of many thousands of teeth of *Australopithecus, Homo erectus*, Neanderthal man and modern man. Nevertheless, cognizance must be taken of the fact that evolution probably involves tooth complexes rather than single teeth (Perzigian, 1975).

Incisors

There were few statistically significant differences between the mean values calculated for the incisor linear dimensions. When australopithecines were compared with *Homo erectus*, the maxillary lateral incisors appeared significantly larger in the latter genus. In turn, when *Homo erectus* was compared with the Neanderthal specimens, the latter had significantly broader central incisors. There was a significant decrease in robustness and buccolingual diameter of the maxillary incisors with the appearance of modern man.

Canines

A statistically significant difference was found in the mesiodistal and buccolingual diameters and robustness in mandibular canines between *Australopithecus africanus* and *Australopithecus robustus,* but neither taxon differed significantly in these dimensions from *Homo erectus.* There were no significant differences in this tooth between *Homo erectus* and Neanderthal man in any measure, but extant *Homo sapiens* showed a significant reduction from the latter in all dimensions.

A comparison of the australopithecine maxillary canines produced no statistically significant differences for the mesiodistal and buccolingual diameters or robustness, nor were any such differences disclosed when the australopithecines were compared as a single taxon with *Homo erectus.* There was a reduction in the mesiodistal diameter and robustness between *Homo erectus* and the Neanderthal specimens and a further reduction between the latter and contemporary *Homo sapiens.*

Premolars

The mandibular first premolars were exceptional in that there were no significant differences in their mesiodistal or buccolingual diameters or robustness between the australopithecine groups. Otherwise, all the premolars varied in the same way. *Australopithecus robustus* had values significantly larger than *Australopithecus africanus* while those for *Homo erectus* were significantly smaller than for either of the australopithecine taxa separately or both taken as a single taxon. The Neanderthal values were significantly smaller than those for *Homo erectus* but there was no significant difference between the former and the values for extant man. In general, there appears to have been a trend towards reduction in premolar size.

Molars

The buccolingual and mesiodistal diameters and robustness of the mandibular molars differed between the australopithecines by amounts that were either not statistically significant or barely so. *Homo erectus,* on the other hand, had significantly smaller mandibular molars than either the australopithecine taxa taken separately or together. The Neanderthal teeth were significantly smaller than those for *Homo erectus,* but they were not significantly different from those of modern man. The maxillary molars showed somewhat parallel differences to those observed in the mandibular teeth.

Variability

The combined data for variability changes showed first a clear trend for increased variability, tending to corroborate the reduced selection hypothesis. In some instances, the greatest variability was found in the Neanderthal teeth, while in others, modern man exhibited maximum variability. This was not surprising, since many teeth of the earlier group had already been reduced to modern size.

Also, Perzigian (1976) reports that the coefficients of variation were greater for the more distal than mesial teeth in the Indian Knoll population, but whether this pertains to other primates has yet to be elucidated. Hence, any interpretation would be premature.

It seems that an evolutionary reduction in hominid tooth size has occurred, although the possible mechanisms remain obscure. One possibility is that tooth reduction is associated with changes in the physical consistency of the diet (Brace, 1962), the assumption being that during human evolution, the habitual diet became progressively softer and required less mastication. The appearance of fire may well have resulted in the suspension of the remaining selection for relatively large posterior teeth, so that, by the Neanderthal stage, the posterior teeth had been reduced to modern size. Jolly (1970), alternatively, has suggested that incisor reduction may be an adaptation to small object feeding and that canine reduction is a secondary but dependent feature with the rotary feeding habit.

There is diversity of opinion concerning the effect of the development of tools and weapons on the evolution of the dentition (Oppenheimer, 1966). There is no way of deciding whether tooth

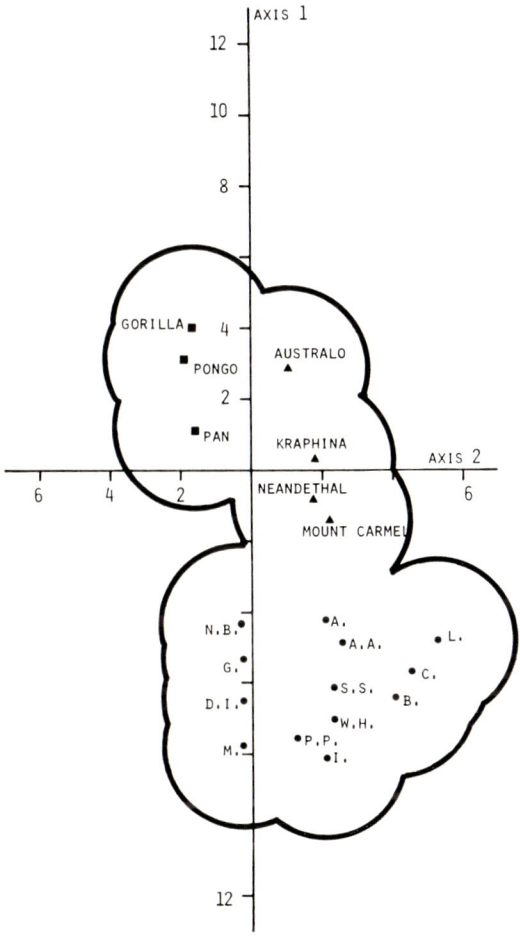

Figure 20. Variation in ape and hominoid tooth size. First two axes of the canonical analysis of the mesiodistal and buccolingual mandibular permanent tooth dimensions combined together: centroids and 90 percent confidence limits. Capital letters represent different population samples of *Homo sapiens*.

reduction preceded and possibly precipitated a need for tool use or whether tool use antedated and possibly determined in some way tooth reduction. The considerable variation in tooth size among the Hominoidea and particularly the observation that fossil hominoids were widespread at the close of the Miocene with

164 *Evolutionary Changes to the Primate Skull and Dentition*

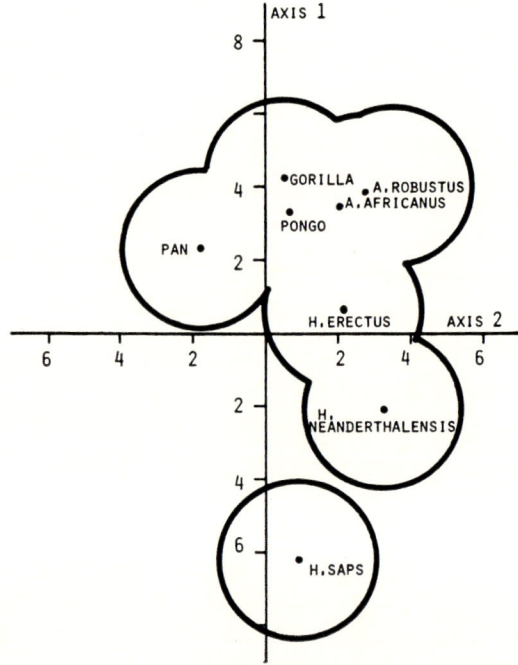

Figure 21. Variation in ape and hominoid tooth size. First two axes of the canonical analysis of the mesiodistal and buccolingual mandibular permanent tooth dimensions combined together: centroids and 90 percent confidence limits. The same population samples were included in this analysis as for Figure 20, although their groupings differed.

anterior teeth much smaller than those of extant apes, suggests caution when directly associating tooth size and tool use.

Another possibility is that the reduction in tooth size is just a part of the overall phylogenetic shortening of the face (Hooton, 1946; Mills, 1963). Because of the better balancing of the head that it confers, the shortening of the facial skeleton is itself often regarded as an inevitable corollary of the adoption of an orthograde posture. There is no a priori reason why this should be so as head poise and head balance in the Hominoidea are two separate, and not necessarily highly correlated, attributes (Moore, Adams and Lavelle, 1973). Elsewhere in the Primates, there is no obvious association between facial length and the adoption of quadrupedalism or brachiation (Walker, 1967).

Yet, an alternative suggestion relates tooth reduction to changes in behavior. Holloway (1967), for instance, has suggested that canine reduction is related to a change in hormonal factors that occurred at some point in the development of the hominid behavior pattern.

The development of hominid dental traits may not have resulted entirely from specific adaptation. Despite the general opinion that drift had not been an evolutionary mechanism of great importance, it cannot be entirely dismissed from a role in the evolution of the hominid dentition, particularly in its later stages. As a result of cultural attainments, *Homo* has probably always undergone extensive population movements. Under circumstances, where the population size is repeatedly fluctuating and occasionally falling to small numbers, drift can become an important contributor to genetic variation (Wright, 1948). This situation is quite unlike that of most other organisms where movement is stimulated only by population pressure and relatively slow environmental fluctuations. Drift and selection are not mutually exclusive; both can affect gene frequencies simultaneously (Dobzhansky and Pavolvsky, 1957).

The frequency of dental variants can also be maintained or changed by pleiotropy in a manner not wholly dependent on the adaptive value of the variant. The evolutionary influence acting on a population depends upon its *average* fitness in a usually fluctuating environment (Lewontin, 1965). The likelihood of such mechanisms being responsible for maintenance of different dental phenotypes is increased by the fact that most dental characteristics are genetically complex. The interaction effects of genes responsible for dental traits may contribute more to the average fitness than the dental traits themselves. Evolutionary interpretations based on dental characteristics must, therefore, account for the fact that teeth are only a small part of a complex organism and are not necessarily the primary adaptive elements.

RATES OF EVOLUTIONARY CHANGE IN THE HOMINID DENTITION ATTRITION

Determination of rates of evolutionary change is fundamental to evolutionary studies, and physical anthropologists have shown increasing interest in the rates of morphological change revealed by the fossil record.

The rates of evolutionary change may be estimated from proportional change per million years, calculated according to the formula of Haldane (1949); rate =

$$\frac{\log_e X_2 - \log_e X_1}{t_2 - t_1}$$

where X_1 and X_2 are the values of the parameter at time t_1 and t_2. The result is expressed in darwins (one darwin is a change of a factor of "e" in one million years). A negative rate implies a reduction in the parameter.

In considering the long-term changes in the hominid premolar and molar dimensions, the functionally significant aspect appears to be the change in crown area available for chewing. Simple estimates of this change may be obtained from the mesiodistal and buccolingual crown diameters.

Using the generally accepted hominid phylogeny, Bilsborough (1969) has calculated the evolutionary rates in hominid premolar and molar dimensions from the following fossil sequence: Olduvai Bed I hominids 4,6,7 (1.75 million years); Olduvai Bed II hominid 13 (1 million years); *Homo erectus* from Java—Pithecanthropus IX maxilla and Sangiran B mandible (700,000 years) and later material from Choukoutien, China (370,000 years). Due to the lack of comprehensive premolar and molar tooth dimensions for *Homo sapiens,* Bilsborough used the midvalues previously published for microdont Bushman (Drennan, 1929) and macrodont Australian aborigines (Campbell, 1925). In the following account, the very small sample size of the fossil specimens must be borne in mind.

The rates of evolutionary change varied from tooth to tooth, nevertheless, some overall trends emerged. In size and to a lesser extent in shape changes, there is considerable, although not complete, symmetry in rates of evolutionary change of both the maxillary and mandibular premolar and molar dentitions; the rates of change between Olduvai Bed I and Bed II and between Olduvai Bed II and *Homo erectus* (Java) were low, i.e. there was a low rate of evolutionary change in Lower Pleistocene; the rates between *Homo erectus* (Java) and *Homo erectus* (Pekin) were very high for the size of the second and third molars, i.e. Middle Pleistocene was a period of maximum change in the molars; between *Homo erectus* (Pekin) and *Homo sapiens*, there were high rates of change in premolar size, while the rates of change in molars were less, also the rates of change of shape were lower and more variable; between Olduvai hominid 13 and *Homo sapiens*, the rates of evolutionary changes in the premolars and second and third molars were similar, but lower rates of change existed for the first molar.

From these unavoidably restricted samples it appears, therefore, that evolution in certain dental parameters was proceeding relatively slowly during the Lower Pleistocene but had speeded up considerably by the middle of the epoch. Earlier studies dealing with the rate of evolutionary change in cranial capacity (Haldane, 1949; Campbell, 1963) have provided evidence that this too was low in the Lower Pleistocene but increased (to about 375 millidarwins) during the Middle Pleistocene. It is possible that the underlying cause of these variations in evolutionary rate shared some common factor, perhaps a change in selection pressure.

The principal difficulty in computing rates of evolution in hominids arises from imprecision in dating the hominid fossils. In *Australopithecus*, for instance, estimates of the dating of Transvaal sites vary from three million years B.P. to 600,000 years B.P., while one of the earliest fragments attributed to this group (the Lothagam jaw) is dated at 5.5 million years B.P. Similar difficulties are encountered in the dating of both *Homo erectus* and Neanderthal fossils.

In addition, Keene (1967) described two unrelated American Negro males with very large teeth. The maximum reported values for the mesiodistal diameter of Australian aboriginals were exceeded in eight out of the sixteen tooth groups, and the majority of the Negro dental measurements either exceeded or lay between the mean values reported in the literature of *Australopithecus* and *Paranthropus*. Keene (1967) interpreted this to cast doubt on the concept that a distinct morphological gap separates fossil from contemporary man. Furthermore, Corruccini (1972) reported that large-toothed human races exhibit a closer relationship to *Australopithecus* than caucasoids. Indeed, the dimensions of caucasoid teeth show a higher similarity to the great apes as a whole than do the more primitive and phylogenetically more ape-like australopithecine teeth. Obviously, therefore, a great deal more data are required before the evolutionary changes in the primate dentition can be elucidated, especially relating to the variability of tooth dimensions in nonhuman primate groups (Hrdlicka, 1923; Colyer, 1943; Schultz, 1958; Swindler, Gavan and Turner, 1963; Biggerstaff, 1966; Swindler, McCoy and Hornbeck, 1967; Leutenegger, 1971; Almquist, 1974; Butler and Bernstein, 1974; Swindler and Orlosky, 1974; Kinzey, 1974; Orlosky, Swindler, and McCoy-Beck, 1974).

Some characteristics of mammals evolve much faster than others which evolve at the standard rate. This dichotomy is quite unexpected, but appears post facto to be predicted by a new and ecological account of evolution.

For instance, the teeth of horses exhibit both modes of evolution. Their measurements evolve in the standard mode, whereas their patterns of enamel, dentine, and cement evolve in the epistandard mode as judged directly and indirectly by the rate of origination and extinction of taxa that are defined in part by their teeth patterning. The difference between the modes clearly is not that of neutral against adaptive evolution. Mensurational aspects of horse teeth are obviously of selective importance. The degree of constancy for size evolution is, nevertheless, indistinguishable now from that of protein evolution, which has some more or less predictable deviations from constancy and which is measured on a much grosser scale than is size evolution.

What is the possible explanation of the epistandard mode? In the resource of space, the proportion of the resources used by one species changes over evolutionary time; usually if one species gains resources others lose. Presumably, the boundaries in the resource space between species shift back and forth rather than there being mostly unidirectional changes. Indeed, the rate of extinction depends on the susceptibility of the species at risk; the greater the proportion of its resources it loses, the greater the risk of extinction.

For mammals, the spectral threshold of ultimately regulatory resources needed to prevent extinction is presumably lower than in most organisms. Mammals become extinct more often than most organisms with similar fluctuations in the partitioning of the resource space. Epistandard origination rates are needed to replace the extinctions; possibly this just involves greater survival of incipient species which could themselves help cause the extinctions. The apparently epistandard evolution of morphological pattern is clearly consistent with this situation, as is the epistandard reshuffling of the genome. The latter effects, however, may not be necessary causal adjuncts.

Why should mammals be so susceptible to extinction or have such a high rate of formations or survivals of incipient species? There are several possible explanations, related to mammalian attributes such as the high rate at which they use energy, their apparently sharply bounded but contiguous adaptive zones, the ease with which their genomes become reorganized or their high-level feedback systems. It is clear, therefore, that phenomena related to organization or extinction rates should participate in the epistandard mode.

For a group whose taxa have a low threshold for extinction, e.g. mammals, therefore, some phenomena will evolve in the epistandard mode in some type of causal relation to this low threshold. Other phenomena will be unaffected by the causal nexus of the threshold and will evolve in relation to the standard losses and gains in resources (realized fitness). It is apparent, therefore, that the modes of evolution are far more complex than hitherto assumed, so that the interpretation of the rates of dental evolutionary change will require considerably more data than currently exists to receive an adequate explanation.

Furthermore, there is little data on how the rates of evolutionary change of the dentition are associated with those of the primate skull. Obviously, the causal factors resulting in dentofacial change will result from both environmental and genetic changes, but how far each is involved requires further study. For instance, from the studies of Lavelle (1974), Hamori, Gyulavari, Szabo (1974), and Suarez and Spence (1974), there are complex associations between tooth and skull size, but how these are reflected in rates of evolutionary change is unknown. Thus, while low correlations between tooth and body size have been computed for American Negroes (Henderson and Corruccini, 1976), high correlations have been reported in *Macaca* (Lauer, 1975).

ATTRITION

Dental attrition is a particularly complex polymorphism, being dependent upon masticatory function, physical consistency of the diet, timing and sequence of tooth eruption, tooth form, tooth position within the dental arcade, thickness and hardness of enamel, and predisposing factors such as enamel dysplasia and habits such as bruxism. Of these many factors, the nature of the diet and the consequent vigor of mastication are assumed to be the most important. Begg (1954) suggested that attrition should be regarded as just the inevitable accompaniment of mastication, but that its occurrence is functionally beneficial in that it prevents locking of the teeth in occlusion and the persistence into adult life of minor degrees of malocclusion produced during growth and eruption.

Attrition is normally regarded only in relation to the permanent teeth, whereas considerable attrition may also occur in the deciduous dentition. Nanda, Khan, and Anand (1973) considered attrition of the deciduous dentition to be important in determining the tooth position and occlusion of the permanent dentition. Attrition of the deciduous dentition appears to have escaped the determination of most workers, however.

Occlusal Attrition

Many qualitative studies of occlusal attrition are based on simple criteria such as minor faceting, dentine exposure at the tips of cusps, and fissure obliteration (Miles, 1963; Butler, 1972; Zanowiak, 1974).

Two main attributes of attrition can be assessed, the pattern of dentine exposure and the difference in wear between adjoining teeth. The pattern of dentine exposure is well documented for both living apes and modern man. The most comprehensive account has been provided by Ashton and Zuckerman (1954), who examined several hundred gorilla, chimpanzee, and modern human skulls. They observed a common pattern of attrition in which wear occurred most on the inner cusps of the maxillary and outer cusps of the mandibular molars. They also noted that the order of appearance of foci of attrition was similar in man and apes, and regarded such findings as indicative of a common pattern of mastication in all three groups. Minor degrees of variation could still exist, however, in the order and appearance of the facets and the rate at which they progress. These might be taxonomically significant by reflecting small but specific differences in the method of chewing or in the nature of the diet.

In order to evaluate such minor differences in attrition and take into account the wide individual variation which undoubtedly exists, some quantification is desirable. For instance, Murphy (1959) compared the patterns of attrition in adjoining teeth. Using such a method, Lavelle (1970) has investigated attrition in various human ethnic groups. It was found that in each group, the degree of attrition increased from M1 to M3, although not infrequently the reverse situation pertained in the maxillary molars. In general, molar attrition was greater in Anglo-Saxon caucasoids, negroids, mongoloids, and australoids than in modern British caucasoids. Dietary habits probably underlie this difference, although attrition is such a complex polymorphism that it cannot be taken for granted that dietary factors are always significant.

Occlusal attrition is generally recognized to result from two distinctly different types of masticatory behavior. Food is initially ingested and then pulped by a series of puncture-crushing masticatory cycles in which no tooth-tooth contact occurs. This activity leads to blunting and ultimately cavitation of the tips of the cusps (Hiiemae and Kay, 1973). When the food is sufficiently softened the occlusal surfaces of the maxillary and mandibular teeth come into close contact. The relative movement of the maxillary and mandibular molars is controlled by their crown morphology. This second type of masticatory cycle leads to striated and gouged facets on the occlusal surfaces of teeth, resulting either from tooth-tooth contact or contact between teeth and abrasive food particles. Recently, rather than regarding this latter tooth wear as "natural attrition," Every (1974) has advanced the hypothesis that attrition is the result of a "phylogenetically induced behavior" termed thegosis (or tooth sharpening). According to this worker, the movements of thegosis are opposite to those of mastication and the striations are produced by small enamel chips gouging the tooth surface during a thegotic movement. Yet, although there is experimental confirmation based on cinefluorographic and occlusal analysis in *Tupaia, Galago, Saimiri,* and *Ateles* (Kay and Hiiemae, 1974), further proof is required before the notion of thegosis can be regarded as a universal primate attribute.

Interproximal Attrition

Attrition at the contact areas between adjacent teeth occurs in most mammalian species. It results in a reduction of the mesiodistal diameter and once the occlusal cusp pattern is abraded, in forward migration of the teeth in the arcade. Thus, the dental arch gradually shortens with age (Goose, 1963). Interproximal attrition results from the lateral motion of adjacent teeth relative to each other in the plane perpendicular to the line between contiguous contact areas. This motion originates from two sources: lateral resolution of vertical occlusal forces due to the lateral angulation shared by all teeth within the arch (Kraus, Jordan, and Abrams, 1969), lateral jaw movements occurring during mastication. The teeth are kept in contact despite the loss of tooth substance from the contact areas by the mesial force vector (Picton, 1962). This vector is said to comprise six major forces: the formation and eruption of permanent molars (Goldstein and Stanton, 1935), mesially directed forces resulting from tongue and cheek movements (Picton, 1962), continued tooth eruption throughout life (Begg, 1954), deposition of bone on the distal aspect and resorption from the mesial aspect of the alveoli (Brash, 1953), force of occlusion of the tooth crowns (Picton, 1964), and components of the occlusal force that act tangentially and anteriorly to the alveolar margin (Wolpoff, 1971). Experimental studies have indicated that interproximal attrition may result from a low level of mesial force (Picton, 1962), and that it is closely related to the mesial tilt of the teeth within the alveolus (Moss and Picton, 1970).

There have been few attempts to quantify the amount of

tooth loss due to this type of attrition. In the classic study of Begg (1954), tooth length was compared in both worn and unworn teeth of *Homo sapiens*, where the tooth loss for individual teeth ranged from 5 percent for second premolars to 18 percent for first molars. Begg concluded that the loss of tooth substance and concomitant forward migration of teeth within the arch once the cusp pattern is lost due to attrition, occurs throughout life. The degree of attrition was probably much greater in fossil groups than in extant populations and probably led to the edge-to-edge incisor occlusion commonly seen in the former.

There are two ways of regarding interproximal attrition. The first is to view it as a source of error which is introduced into species description, although if only teeth with no attrition were described, then there would be few teeth from fossil specimens. The second way is to regard attrition as a "normal" physiological event, so that ideally in any comparison, samples should comprise both worn and unworn teeth.

ERUPTION SEQUENCE

A discussion of the mechanism of tooth eruption is beyond the scope of this chapter, and the reader is referred to reviews by Schour (1934), and Ness (1964) and to published investigations by Thomas (1964) and Main and Adams (1966).

The deciduous tooth eruption sequence in *Homo sapiens* has been the subject of many investigations (e.g. Niswander and Sujaku, 1960; Friedlander and Bailit, 1969). The sequence of eruption with approximate chronological ages is incisors (6-9 months), first molars (12-14 months), and second molars (24-30 months). There is considerable individual variation(Lysell, Magnussen, and Thilander, 1962), but the average sequence appears to be relatively constant between the different ethnic groups (Leighton and Orth, 1968). The eruption sequence has received much less attention in nonhuman primates. The limited evidence that is available indicates that the eruption sequence in the apes is similar to that in man, except that the second molar precedes the canine (Missen and Rissen, 1945; Hurme and Van Waganen, 1953).

The permanent tooth eruption sequence is well documented

for most primate groups. In general, estimates for living man are based on the eruption through the gum, whereas for most other primate genera, the eruption sequence is based upon the order through the bony alveolus, as in fossil specimens. Garn and Lewis (1957) reported that the eruption sequence through the alveolus does not necessarily correlate with that through the gum. A further difficulty arises from the age at which teeth erupt, since this is subject to great variability.

The eruption of the permanent dentition in contemporary human populations has been the subject of a very large number of enquiries (e.g. Hellman, 1923; Hurme, 1948; Gates, 1964; Knott and Meredith, 1966; Alstad, 1973; Perreault, Demirjian and Jenicek, 1974) from which it appears that the most frequent sequence in the upper jaw is M1, I1, I2, P1, P2, C, M2, M3, and in the lower jaw M1, I1, I2, C, P1, P2, M2, M3. Considerable variation exists in the sequence as well as the timing of eruption (Brues, 1958; Moorrees, 1965), with the M1I1 and C,P1P2 sequences being particularly variable (Schultz, 1940). This variability has many causes, including premature loss of deciduous teeth (Miller, Hobsen, and Gaskell, 1965), environmental and nutritional factors (Garn and Russell, 1971), socioeconomic factors (Clements, Davies-Thomas, and Pickett, 1957), variation in fluoride intake (Tank and Strovick, 1964) and systemic disease (Hurme, 1948). Garn and Lewis (1957) divided tooth development into a number of stages, and concluded that a given order of tooth development at one stage is of relatively poor predictive value in estimating the sequence at a subsequent stage. This emphasizes the necessity of comparing eruption sequences at the same developmental stage. Furthermore, Garn and Lewis (1963) report that following population studies, the eruption sequence of P2M2/M2P2 may be determined genetically, although there is yet little supportive evidence, with the exception of the third molar (Garn, Lewis, and Bonne, 1961).

The sequence of permanent tooth eruption shows a marked shift through the order Primates. At one extreme in insectivores, the permanent molars erupt before the deciduous anterior teeth commence replacement; at the other extreme, typified by *Homo sapiens,* the deciduous anterior teeth are largely replaced by their permanent successors before the second and third molars erupt into the dental arcade. The primate data are, however, far from complete. Zuckerman (1928), Clements and Zuckerman (1953), and Schultz (1935) have published extensive records on eruption in *Gorilla* and *Pan*, and Baume and Becks (1950) have examined the development of the dentition in *Macaca mulatta*. Weidenreich (1937) has reviewed eruption in European neanderthaloids and Broom and Robinson (1951) have predicted the "likely" eruption sequence in australopithecine fossil fragments. It must be emphasized, however, that the data are far too scant to permit taxonomic values to be assigned to eruption sequences.

VARIATION IN TOOTH NUMBER

The majority of inherited malformations, including variations in tooth number, are believed to be determined by polygenic or multifactorial inheritance. Such abnormalities have the following characteristics: the condition is relatively common; there are family aggregates; the recurrence rate in siblings is 1 to 5 percent; there is evidence of response to factors such as parental age, birth rank, and sex. Carter (1969) has constructed a model of polygenic inheritance in which he postulated that the genetic predisposition to a particular malformation is distributed normally within the population and that a threshold exists which, when passed, leads to expression of the defect. The level of the threshold may be shifted by environmental factors. The precise mechanism by which genetic predispositions act and how they are influenced by environmental factors in man are not understood. In experimental animals, however, Gruneberg (1953) has demonstrated that in the discontinuous variant "absence of third molars" seen in inbred mouse strains, there is an underlying continuous variation involving tooth size. In those strains in which tooth size falls below a certain threshold, there is a tendency for the tooth to be missing altogether. Grewal (1962) observed that the tooth germs failed to progress beyond the early cap stage in affected animals, suggesting that the basic factor underlying the absence of third molars is the size of the tooth germ at some critical stage of development. Whether this also applies to primates has yet to be established, and there is no supportive evidence that agenesis results from systemic (Grahnen, 1956) or endocrinological (Howard, 1926) factors. The general consensus of opinion appears to support the view that the tendency to tooth agenesis in *Homo sapiens* is primarily due to genetic factors (Watson and Hardwick, 1971; Hume, 1972), but whether these are recessive (Iltis, 1948) or dominant (Montague, 1940) is still uncertain.

The incidence of supernumerary teeth (polygenesis or polydontia) and congenitally missing teeth (anodontia, ologodontia, or agenesis) in the Anthropoidea has been the subject of considerable but scattered literature. The majority of studies have dealt with these conditions in the human dentition, although the contribution of Colyer (1936) provided a wide survey of dental anomalies in numerous genera of monkey and ape. The early literature especially abounds with reports of missing or supernumerary teeth which generally treat variation in tooth number as a phenomenon. Except for highly inbred, hybrid, or trihybrid populations (Thomsen, 1952), where an increased proportion of variation in tooth number may be expected (Grahnen, 1962), clinical rather than theoretical reasons appear to have dictated the study of variation in tooth number. Little attention has been given to the possibility that tooth polymorphisms in man are not isolated phenomena but are fundamentally related to the size, development and calcification timing of the dentition as a whole or the mechanisms that must maintain tooth number polymorphisms in man.

Variations in tooth number are not uncommon. For example, the prevalence of agenesis in the most variable teeth (third molars, lateral incisors, and second molars) ranges between 1 to 30 percent (Garn et al., 1963). Cases of anodontia in the maxillary and mandibular permanent dentitions have been described by Hutchison (1953), and of oligodontia by Brodie and Sarnat (1942). Both total anodontia and the severer forms of oligodontia are much rarer defects than the minor degrees of agenesis, usually occurring as part of a genetically determined systemic anomaly characterized also by the absence of sweat glands (McDonald, 1949), and is frequently associated with a generalized delay in dental development (Bailit and Sung, 1968).

Recent surveys of the prevalence of agenesis in the maxillary and mandibular permanent dentitions have been made by Clayton (1956) and the subject has been reviewed by Montague (1940). From these sources, agenesis in the human dentition appears to occur most frequently in the maxillary incisor region, while maxillary and mandibular third-molar agenesis occur somewhat less commonly. The latter condition has been described in the four major groups: caucasoids (Fanning, 1962); negroids (Chagula, 1960); mongoloids (Goldstein, 1932); australoids (Campbell, 1925). It is often associated with an increased tendency for agenesis of other members of the dentition, with a disturbance of the tooth eruption sequence (Garn, Lewis and Vicinus, 1962), increased morphological variability in the dentition (Keene, 1965), and reduction in both arch size and the size of the other teeth (Lavelle, Ashton and Flinn, 1970), but a compensatory increase in the size of the teeth on the opposite side (Sofaer, Chung, Niswander and Runk, 1971). It is commoner in women than in men. From a study of a large series of cases of agenesis of one or more teeth, Baum and Cohn (1971) concluded that the first molar and canine teeth are the most stable teeth in the human dentition.

Purely visual examination, whether of skulls or patients, can lead to overestimates of the true frequency of occurrence, since unerupted teeth may be counted as absent. Radiographic studies are essential if forming or unerupted teeth are to be detected. Since third-molar calcification in man may not be apparent radiographically until the teens (Banks, 1934), the exclusion of immature specimens or subjects should increase the precision of estimates.

Recently, attention has been directed to the individual correlates of hypodontia or agenesis. Third-molar agenesis is associated with a thirteen-fold increase in the prevalence of other missing teeth (Garn and Lewis, 1962) and with reduction in cusp number and other alterations in crown morphology (Garn and Lewis, 1962; Keene, 1965). It may also be associated with crown size reduction of the remaining teeth (Keene, 1964) and with delayed somatic development and developmental immaturity at birth (Bailit and Sung, 1968). These findings suggest three possibilities: that agenesis has a progressive or incremental effect upon crown size reduction in all the remaining teeth, that there is a gradient of crown size reduction in agenesis from the most posterior teeth to the midline, and that agenesis is associated with a distinctive alteration in the crown-size profile pattern of the remaining teeth, such that even minimal agenesis is reflected in a systematic departure from the usual.

Garn and Lewis (1970) have examined these possibilities by measuring the mesiodistal diameters of the teeth of 658 (mostly Northwestern European) subjects, of which eighty-two had radiographically confirmed agenesis of one or more third-molar teeth and a further nineteen possessed agenesis of multiple teeth, usually lateral incisors or second premolars. Analysis of the data confirmed that agenesis is simply associated with crown-size reduction, and that this reduction is more marked in cases of multiple-tooth agenesis than with third-molar agenesis alone. Agenesis and multiple agenesis were characterized first by size reduction of the remaining teeth and second by variations in size variance of the teeth that are still present and erupted. The results also indicated that there is a gradient of reduction in agenesis, such that the anterior teeth are most reduced and the posterior teeth least reduced.

Third-molar agenesis is associated with delayed formation timing of all of the remaining teeth, especially for the most posterior teeth, and the distal tooth of each class (Bailit and Sung, 1968). There is also an association with reductions in the cusp number of the molars (Lavelle, Ashton and Flinn, 1970), and therefore with the relative number of cusps of M1 and M2 (Davies, 1968), with crown-size reduction, particularly of the lateral incisors, second premolars and molars, and with hypoplastic or "peg-shaped" lateral incisors (Garn et al., 1964). It thus appears that third-molar agenesis has multiple dental and somatic correlates. Either alone or in combination with other genes affecting growth and development, third-molar number polymorphisms may be indicative of multiple developmental delays.

Two gradients appear to be recognizable in third-molar agenesis: (a) a distal-mesial gradient of tooth formation delay, and (b) a mesial-distal gradient of relative size reductions. The general population could, therefore, be considered as two phenotypic groups, namely the agenesis group with timing gradients and size reduction gradients operating in opposite directions and the "normal" group lacking both of these gradients. The variation in the number of each of these two groups may account for the ethnic variations reported in the literature.

Supernumerary teeth occur with a much lower frequency than cases of agenesis in modern man (Diamond, 1935). The most frequent site for this anomaly is in the maxillary central and lateral incisor regions (Stafne, 1931), although maxillary and mandibular supernumeraries have been reported also in the canine (Stafne, 1931), premolar (Cowan, 1952), and third-molar (Peak, 1967) regions. Discrepancies in the incidence of supernumerary teeth between reports of different research workers have probably been enhanced by the high proportion of supernumerary teeth which fail to erupt (Morgan, 1946) and by the great variation in supernumerary tooth form (Howard, 1967) which may obscure diagnosis. In addition, the prevalence of these teeth tends to be greater in cases of congenital malformations of the face, e.g. cleft lip and palate (Millhon and Stafne, 1941), whose inclusion in a survey depends upon sampling procedures. Occasionally, supernumerary and congenitally missing teeth are found in the same individual (Camilleri, 1967).

In an attempt to provide comparative data based upon radiographic examination, Lavelle and Moore (1973) determined the incidence of dental agenesis and polygenesis in a large number of Old World monkeys, apes and caucasoid and negroid representatives of *Homo sapiens* (Tables XVI and XVII). The frequency of agenesis was highest in the human sample, particularly in negroids, while that of polygenesis was highest in great apes.

The average total incidence of dental agenesis was similar in the great apes (0.0 percent in maxilla; 1.0 percent in mandible), lesser apes (0.0 and 1.0 percent) and the majority of the Old World monkeys (averaging 1.2 and 0.6 percent) and much less than in man (averaging for the two ethnic groups 5.2 and 5.9 percent). In contrast, the incidence of polygenesis was higher in

TABLE XVI
TOTAL INCIDENCE OF AGENESIS AND POLYGENESIS IN THE PRIMATE DENTITION

	Number of Specimens	AGENESIS			POLYGENESIS		
		Maxilla	Mandible	Total	Maxilla	Mandible	Total
Cercopithecoidea	978	1.2	0.6	1.8	0.2	0.5	0.7
Hominoidea (a) Hylobatinae	194	–	1.0	1.0	2.1	–	2.1
(b) Ponginae	390	–	1.0	1.0	2.3	3.8	6.2
(c) Hominidae	5000	5.2	5.9	11.1	1.2	0.7	1.9
Total for Hominoidea	5584	4.7	5.4	10.1	1.3	0.9	2.2
Hominoidea and Cercopithecoidea Total	6562	4.2	4.7	8.9	1.1	0.9	2.0

Cercopithecoidea comprized *Colobus, Presbytis, Cercopithecus, Macaca* and *Papio*.
Hylobatinae comprized *Hylobates* and *Symphylangus*.
Ponginae comprized *Gorilla, Pan* and *Pongo*.
Hominidae comprized caucasoids and negroids.

TABLE XVII
PERCENTAGE DISTRIBUTION OF AGENESIS AND POLYGENESIS

	AGENESIS				POLYGENESIS			
	Incisors	Premolars	Molars	Total	Incisors	Premolars	Molars	Total
Cercopithecoidea	0.3	0.6	0.9	1.8	0.1	–	0.6	0.7
Hominoidea (a) Hylobatinae	0.5	0.5	–	1.0	–	1.0	1.0	2.1
(b) Ponginae	–	0.5	0.5	1.0	1.0	–	5.1	6.2
(c) Hominidae	0.7	0.3	10.1	11.1	0.8	0.7	0.4	1.9
Total for Hominoidea	0.7	0.3	9.1	10.1	0.8	0.6	0.7	2.2
Hominoidea and Cercopithecoidea Total	0.6	0.4	7.9	8.9	0.7	0.5	0.7	2.0

Anomalies in the molar teeth were virtually all in the third molar region.

No anomalies were detected in the canine teeth.

the great apes (2.3 and 3.8 percent) than in the lesser apes (2.1 and 0.0 percent) man (1.2 and 0.7 percent) or monkeys (0.2 and 0.5 percent).

In the breakdown of the distribution of agenesis region by region, a scattered pattern was found in the Cercopithecoidea. The molar region was most frequently affected in *Colobus* (4.3 percent) and *Cercopithecus* (0.9 percent). *Colobus* also suffered a relatively high incidence of agenesis in the premolar region (3.6 percent). In the small sample of *Papio*, agenesis was found in two specimens only (5.3 percent), both in the incisor region. Among the apes, the incidence of agenesis was low in all regions.

In man, the incidence of agenesis was significantly greater in the molar (10.1 percent—predominantly in the third molar) region than in either of the premolar (0.3 percent) or incisor (0.7 percent) regions, a difference more marked in the negroid than caucasoid sample. The total average degree of agenesis for negroids averaged 28.2 percent but only 0.9 percent for caucasoids.

In general, agenesis in the third-molar region was approximately equally distributed in both maxilla and mandible, whereas agenesis in the incisor region was confined to the maxilla and agenesis in the premolar region to the mandible.

Combining the occurrence of polygenesis for both jaws, the incidence of polygenesis was found to average 0.7 percent for Old World monkeys, 2.1 percent for lesser apes, 6.2 percent for great apes, and 1.9 percent for man. The major exceptions included the much lower incidence of polygenesis in *Colobus* and also both ethnic groups of Man where the regional differences found in agenesis were reversed, with the molar regions now showing the lowest (statistically significant) frequency of polymorphism. In the great apes, polygenesis was much commoner than agenesis, particularly in the molar region where its incidence was significantly greater than among the incisor teeth in *Gorilla* and *Pongo*.

The interpretation of these results is difficult, although the high incidence of agenesis seen in man suggests that this may be a phylogenetic accompaniment of the reduction of the maxillomandibular skeleton, which is such a marked feature of human evolution.

Because of the different criteria of assessment, i.e. radiographic, not simple observations, comparison between these data and those from previous studies must be treated cautiously. Nonetheless, the findings relating to both the greater and lesser apes are in close agreement with those given by Colyer (1936). The findings for maxillary incisor agenesis in modern man are broadly similar to those previously given by Mandeville (1949) for caucasoid and Hrdlicka (1921) for negroid groups. Discrepancies also exist between the incidence of agenesis in the molar regions in both human groups determined in this study and in those of other workers. Grahen, for example, found the incidence of third-molar agenesis in caucasoids to be 25 percent while Hellman (1928) found an incidence in negroids of 11 percent—the respective figures for the samples of Moore and Lavelle being 6.5 and 24.7 percent. These discrepancies may be due to sampling differences, varying degrees of homogeneity within the various ethnic groups or the lack of radiographic confirmation in the earlier studies. In addition, in some studies it is impossible to determine from the presentation of the data whether there is one or more third molars missing in each subject.

The full interpretation of these findings would require a knowledge of the aetiology of agenesis but, as already discussed, this is still far from clear. The condition has been associated phylogenetically (e.g. Hooton, 1946) with reduction in the relative size of the dentition which has accompanied the shortening of the facial skeleton during hominid evolution. The finding in the study by Lavelle and Moore that the incidence of tooth agenesis is much greater in *Homo sapiens* than in any of the other primate species might be taken to indicate that this is so. The fact that it is the maxillary or mandibular third molar which is frequently missing fits well with the observations that this tooth is the most variable in its morphology (Nanda, 1960) and is the one most commonly unerupted in the reduced jaws of *Homo sapiens*. The reason for the marked differences between the two human ethnic groups, which is particularly pronounced in the molar region, is unknown, but as previously mentioned, this finding is at variance with the results in some other studies.

That tooth number and jaw size are not necessarily closely correlated is illustrated by the fact that maxillary and mandibular third-molar agenesis is more frequent in large-jawed Eskimos than in smaller-jawed caucasoids (Goldstein, 1932). This suggests that oligodontia and polydontia may be just an example of numerical variability having no special relationship to the phylogenetic development of the jaws. If this were true, however, it might be expected that agenesis and polygenesis would be about equal in incidence. In the human dentition this is not so—agenesis is much commoner than polygenesis—a contrast given added significance by a comparison with great apes which, despite their close taxonomic affiliation with man, have a higher incidence of molar polygenesis than agenesis. On this evidence, it seems possible that any further evolutionary reduction in human jaw size might be accompanied by an increasing incidence of agenesis, especially of the third molar tooth.

However, to say that agenesis is polygenic implies that it is conditioned by a large number of loci, although the exact number of loci is debatable. Moreover, to conclude that agenesis is determined by a large number of loci with additive effects is not to say that affected individuals can arise only through such a mechanism (Guardia, 1973; Suarez and Spence, 1974). A percentage of affected individuals may represent sporadic cases, since there is evidence that missing teeth in some individual represents part of a larger syndrome or may even be determined entirely by environmental factors. It is well known that agenesis is not an isolated phenomenon, but is linked to a complex of other dental changes, e.g. agenesis is related to the frequency of other missing teeth, the size of the remaining teeth and the overall rate of dental development. Thus, although the genetic causology of agenesis may be polygenic rather than simple Mendelian, much more research is required to elucidate the detailed genetic control before any evolutionary significance can be abstracted.

CONCLUSION

Although there is a considerable body of data relating to primate teeth, much more quantitative information is required before the taxonomic significance of primate teeth can be assessed objectively. This is stressed from measurements of the posterior tooth occlusal areas for a sample of Greek and Turkish populations spanning a period of 8500 years (Le Blanc and Black, 1974). Rates of reduction of 2 percent per 1000 years for the maxillary and 1 percent per 1000 years for the mandibular teeth were computed, although the significance of such rates cannot yet be elucidated.

REFERENCES

Adloff, P.: *Das Gebiss des Menschen und der Athropormorphen.* Berlin, Springer, 1908.

Almquist, A. J.: Sexual differences in the anterior dentition in African primates. *Am J Phys Anthropol, 40*:359, 1974.

Alstad, S.: Age variation in dental eruption in Norwegian children. *Norske Tannlaegeforen Tid, 83*:42, 1973.

Ashton, E. H. and Zuckerman, S.: Some quantitative dental characteristics of the chimpanzee, gorilla and orang-utan. *Philos Trans R Soc Lond* (Biol Sci) *234*:471, 1950.

Ashton, E. H. and Zuckerman, S.: The anatomy of the articular fossa (fossa mandibularis) in man and apes. *Am J Phys Anthropol, 12*:29, 1954.

Ashton, E. H., Healy, M. J. R., and Lipton, S.: The descriptive use of discriminant functions in physical anthropology. *Proc R Soc Lond* (Biol), *146*:552, 1957.

Bailit, H. L. and Sung, B.: Maternal effects on the developing dentition. *Archs Oral Biol, 13*:155, 1968.

Banks, H. V.: The incidence of third molar development. *Angle Orthod, 4*: 223, 1934.

Bateson, W.: *Materials for the Study of Variation, treated with especial regard to Discontinuity in the Origin of Species.* London, Bateson Press, 1894.

Baume, L. J. and Becks, H.: The development of the dentition of *Macaca mulatta. Am J Orthod Oral Surg, 36*:723, 1950.

Baum, L. and Cohen, M. M.: Studies on agenesis in the permanent dentition. *Am J Phys Anthropol, 35*:125, 1971.

Bay, R.: Das Gebis has Neanderthalens. In Koenigswald, von G. H. R.: *Hundert Jahre Neanderthaler.* Utrecht, Kenick en Zoon, 1958, pp. 123-140.

Begg, P. R.: Stone-age man's dentition. *Am J Orthod, 40*:298, 373, 462, 517, 1954.

Biggerstaff, R. H.: Metric and taxonomic variation in the dentition of two Asian cereopithecine species: *Macaca malatta* and *Macaca spenciosa. Am J Phys Anthropol, 24*:381, 1966.

Biggerstaff, R. H.: On the groove configuration of mandibular molars: the unreliability of the "Dryopithecus" pattern and a new method for classifying mandibular morals. *Am J Phys Anthropol, 29*:441, 1968.

Biggerstaff, R. H.: Electronic methods for the analysis of the human post-canine dentition. *Am J Phys Anthropol, 31*:163, 1969.

Biggerstaff, R. H.: Hereditability of the Carabelli cusp in twins. *J Dent Res, 52*:40, 1973.

Biggerstaff, R. H.: Cusp size, sexual dimorphism and heritability of cusp size in twins. *Am J Phys Anthropol, 42*:1247, 1975.

Bilsborough, A.: Rate of evolutionary change in the hominid dentition. *Nature* (Lond), *223*:146, 1969.

Blanco, R. and Chakraborty, R. The genetics of shovel shape in maxillary central incisors in man. *Am J Phys Anthropol, 44*:233, 1976.

Blumberg, J. E., Hylander, W. L., and Goop, R. A.: Taurodontism: a biometric study. *Am J Phys Anthropol, 34*:243, 1971.

Bolk, L.: Odontological essays, 4. On the relation between reptilian and mammalian teeth. *J Anat, 56*:107, 1922.

Bolton, W. A.: Disharmony in tooth size and its relation to the analysis and treatment of malocclusion. *Angle Orthod, 28*:113, 1958.

Bowden, D. E. J. and Goose, D. H.: Inheritance of tooth size in Liverpool families. *J Med Genet, 6*:55, 1969.

Brace, C. L.: Cultural factors in the evolution of the human dentition. In Montagu, M.F.A. (Ed.): *Culture and the Evolution of Man*. New York, Oxford U Pr, 1962, pp. 343-354.

Brace, C. L.: *The Stages of Evolution*. Englewood Cliffs, Prentice Hall, 1967a.

Brace, C. L.: Environment, tooth form and size in the Pleistocene. *J Dent Res, 46*:809, 1967b.

Brace, C. L.: More on the fate of the "classic" Neanderthals. *Curr Anthropol, 7*:204, 1967c.

Brace, C. L. and Montague, M.F.A.: *Man's Evolution*. New York, Macmillan, 1965.

Brash, J. C.: Comparative anatomy of tooth movement during growth of the jaws. *Dent Rec, 73*:460, 1953.

Brodie, A. G. and Sarnat, B. G.: Ectodermal dysplasia with complete anodontia. *Am J Dis Child, 64*:1046, 1942.

Broom, R and Robinson, J. T.: Eruption of the permanent teeth in the South African fossil ape-man. *Nature* (Lond), *167*:443, 1951.

Brues, A. M.: Variation in individual tooth development. *J Crim Law, Criminology and Police Science, 48*:551, 1958.

Butler, D. J.: The eruption of teeth and its association with early loss of the deciduous teeth. *Br Dent J, 112*:443, 1962.

Butler, P. M.: Studies of the mammalian dentition, differentiation of the post-canine dentition. *Proc Zool Soc Lond, 109*:1, 1939.

Butler, P. M.: The ontogeny of molar pattern. *Biol Rev, 31*:30, 1956.

Butler, P. M.: The prenatal development of the human first upper permanent molar. *Archs Oral Biol, 12*:551, 1967a.

Butler, P. M.: Relative growth within the human first upper permanent molar during the prenatal period. *Archs Oral Biol, 12*:983, 1967b.

Butler, P. M.: Growth of the human second lower deciduous molar. *Archs Oral Biol, 13*:671, 1968.

Butler, P. M.: Growth of human tooth germs. In Dahlberg, A. A. (Ed.): *Dental Morphology and Evolution*. Chicago, Chicago U Pr, 1971, pp. 3-14.

Butler, P. M. and Mills, J. R. E.: A contribution of the odontology of Oreopithecus. *Bull Br Mus Nat Hist, 4*:1, 1959.

Butler, P. M. and Ramadan, A.: Distribution of mitoses in the human inner enamel epithelium of molar tooth germs of the mouse. *J Dent Res, 41*:1261, 1962.

Butler, P. M.: Some functional aspects of molar evolution. *Evolution, 26*:474, 1972.

Butler, R. J. and Bernstein, I. S.: Canine role in dental wear patterns. *Am J Phys Anthropol, 40*:391, 1974.

Campbell, B. G.: Quantitative taxonomy and human evolution. In Washburn, S. L. (Ed.): *Classification and Human Evolution*. Adelaide, Hassell Press, 1963, pp. 1-123.

Campbell, T. D.: *Dentition and Palate of the Australian Aboriginal*. Adelaide, Hassell Press, 1925.

Carabelli, G.: *Systematishes Handbuch der Zahnheilkunde*. Wein, Braumuller und Siedel, 1842.

Carr, H. G.: Some dental characteristics of the Middle Minoans. *Man, 60*:119, 1960.

Carter, C. O.: The genetics of congenital malformations. *Proc R Soc Med, 61*:991, 1969.

Chagula, W. K.: The age of eruption of the permanent molars in male East Africans. *Am J Phys Anthropol, 18*:77, 1960.

Chase, C. E. and Swindler, D. R.: A system for automatic recording of odontometric material. *J Dent Res, 53*:1506, 1974.

Clalk, W. E. Le Gros.: *Man-apes or Ape-men?* New York, HRW, 1967.

Clark, C., Lavelle, C. L. B., and Flinn, R. M.: Photogrammetric methods of tooth definition. *J Anat, 109*:344, 1971.

Clayton, J. M.: Congenital dental anomalies occurring in 3557 children. *J Dent Child, 23*:206, 1956.

Clements, E. M. B., Davies-Thomas, E., and Pickett, K. G.: Time of eruption of permanent teeth in British children at independent, rural and urban schools. *Br Med J, 1*:1511, 1957.

Clements, E. M. B. and Zuckerman, S.: The order of eruption of the permanent teeth in the Hominoidea. *Am J Phys Anthropol, 11*:313, 1953.

Colyer, F.: *Variations and Diseases of the Teeth of Animals*. London, John Bale, Sons and Danielsson, 1936.

Colyer, F.: Variations of the teeth of the *Colobus* monkey. *Dent Rec, 63*: 109, 134, 159, 183, 206, 232, 1943.

Corruccini, R.: Allometry correction in taximetrics. *Systematic Zool, 21*:375, 1972.

Cowan, G. A.: Delayed development of supernumerary premolars. *Br Dent J, 92*:126, 1952.

Dahlberg, A. A.: The changing dentition of man. *J Am Dent Assoc, 32*:676, 1945.

Dahlberg, A. A.: *The Physical Anthropology of the American Indian*. New York, Viking Fund, 1948.

Dahlberg, A. A.: Dental evolution and culture *Hum Biol, 35*:237, 1963.

Dart, R. A.: The adolescent mandible of *Australopithecus prometheus*. *Am J Phys Anthropol, 6*:391, 1948.

Davies, P. L.: The relationship of cusp reduction in the permanent mandibular first molar to agenesis of teeth. *J Dent Res, 47*:499, 1968.

De Jonge Cohen, Th.: *Anatomie der Zahne*. Munich und Berlin, Zahn-Mund-Kieferhkd, Verlag von Urban u Schwarzenberg, 1958.

De Terra, M.: Beitrage zu einer Odontographie der Menschenrassen. Zurich, University of Zurich, 1905.

Diamond, M.: *Dental Anatomy*. New York, Macmillan, 1935.

Dobzansky, T. and Pavolvski, O.: An experimental study of interaction between genetic drift and natural selection. *Evolution, 11*:311, 1957.

Drennan, M. R.: The dentition of a Bushman tribe. *Ann S Afr Mus, 24*: 61, 1929.

Every, R. G.: Sharpness of teeth in man and other primates. *Postilla, 143*: 1, 1970.

Every, R.: *Thegosis in Presimians*. London, Duckworth, 1974.

Fanning, E. A.: Effect of extraction of deciduous molars of the formation and eruption of their successors. *Angle Orthod, 32*:44. 1962.

Friedlander, J. S. and Bailit, H. L.: Eruption times of the deciduous and permanent teeth of natives of Bongainville Island Territory of New Guinea: a study of racial variation. *Hum Biol, 41*:51, 1969.

Frisch, J. E.: Sex differences in the canines of the gibbon (Hylobateslar). *Primates, 4*:1, 1963.

Frisch, J. E.: Trends in the evolution of the hominoid dentition. *Biblio Primat Fasc, 3*:1, 1965.

Gabriel, A. C.: The correlation of the size of human teeth with one another and with certain jaw measurements. *Dent J Australia, 27*:174, 1955.

Garn, S. M., Koski, K., and Lewis, A. B.: Problems of determining the tooth eruption sequence in fossil and modern man. *Am J Phys Anthropol, 15*: 313, 1957.

Garn, S. M. and Lewis, A. B.: Relationship between the sequence of calcification and the sequence of eruption of the mandibular molar and premolar teeth. *J Dent Res, 36*:992, 1957.

Garn, S. M. and Lewis, A. B.: The relationship between third molar agenesis and reduction in tooth number. *Angle Orthod, 32*:14, 1962.

Garn, S. M. and Lewis, A. B.: Phylogenetic and intra-specific variation in tooth sequence polymorphism. In Brothwell, D. R. (Ed.): *Dental Anthropology.* New York, Pergamon, 1963, pp. 29-52.

Garn, S. M. and Lewis, A. B.: The gradient and the pattern of crown size reduction in simple hypodontia. *Angle Orthod, 40*:51, 1970.

Garn, S. M., Lewis, A. B., and Bonne, B.: Third molar polymorphism and the timing of tooth formation. *Nature* (Lond), *192*:989, 1961.

Garn, S. M., Lewis, A. B., and Kerewsky, R.: Genetic, nutritional and maturational correlates of dental development. *J Dent Res, 44*:228, 1965.

Garn, S. M., Lewis, A. B., and Kerewsky, R.: Buccolingual size asymmetry and its developmental meaning. *Angle Orthod, 37*:186, 1967.

Garn, S. M., Lewis, A. B., and Kerewsky, R.: Relationship between buccolingual and mesoidistal tooth diameters. *J Dent Res, 44*:495, 1968.

Garn, S. M., Lewis, A. B., and Vicinus, J.: Third molar agenesis and reduction in number of other teeth. *J Dent Res, 41*:717, 1962.

Garn, S. M., Lewis, A. B., and Vicinus, J.: Third molar agenesis and its significance to dental genetics. *J Dent Res, 42*:344, 1963.

Garn, S. M., Lewis, A. B., and Vicinus, J.: Molar size sequences and fossil taxonomy. *Dent Abs, 9*:475, 1964.

Garn, S. M., Lewis, A. B., and Walenga, A. J.: Evidence for a secular trend in tooth size over two generations. *Archs Oral Biol, 13*:841, 1968.

Garn, S. M. and Russell, A. L.: The effect of nutritional extremes on dental development. *Clin Nutr, 24*:285, 1971.

Gates, R. E.: Eruption of permanent teeth in New South Wales schoolchildren. Part 1, Age of eruption. *Australian Dent J, 9*:211, 1964.

Gaunt, W. A.: The development of the deciduous cheek teeth of the rat. *Acta Anat, 38*:187, 1959.

Goldstein, M. S.: Congenital absence and impaction of the third molar in the Eskimo mandible. *Am J Phys Anthropol, 16*:381, 1932.

Goldstein, M. S. and Stanton, F. L.: Changes in dimension and form of dental arches with age. *Int J Orthod, 21*:357, 1935.

Goose, D. H.: Variability of the form of maxillary permanent incisors. *J Dent Res, 35*:902, 1956.

Goose, D. H.: Dental measurement; an assessment of its value in anthropological studies. In Brothwell, D. R. (Ed.): *Dental Anthropology.* New York, Pergamon, 1963, pp. 125-148.

Goose, D. H. and Lee, G. T. R.: Inheritance of tooth size in immigrant populations *J Dent Res, 52*:175, 1973.

Grahnen, H.: Hypodontia in the permanent dentition. *Odont Revy, 7*:5, 1956.

Grahnen, H.: Hereditary factors in relation to dental caries and congenitally missing teeth. In Witkop, C. J. (Ed.): *Genetics and Dental Health.* New York, McGraw, 1962, pp. 194-204.

Gregory, W. K.: *The Origin and Evolution of the Human Dentition.* Baltimore, Williams and Wilkins, 1922.

Gregory, W. K. and Hellman, M.: Dentition of the extinct South African apeman, *Plesianthropus. Annals Trans Mus, 19:*339, 1939.

Gregory, W. K., Hellman, M., and Lewis, G. E.: Fossil Anthropoids of the Yale-Cambridge Indian Expedition 1935. *Washington, Carnegie Inst Washington, Publ,* 495, 1936.

Grewal, M. S.: The development of an inherited tooth defect in the mouse. *J Embryol Exp Morphol, 10:*202, 1962.

Gruneberg, H.: *The Pathology of Development.* New York, Wiley, 1953.

Guardia, D. L.: Etilogy of dental agenesis. *Rev Orthop Dento Fac, 7:*17, 1973.

Haldane, J. B. S.: Suggestions as to quantitative measurement of rates of evolution. *Evolution, 3:*51, 1949.

Hamori, I., Gyulavari, O., and Szabo, B.: Tooth size in pituitary dwarfs. *J Dent Res, 53:*1302, 1974.

Hellman, M.: Nutrition, growth and dentition. *Dent Cosmos, 65:*34, 1923.

Hellman, M.: Racial characters in the human dentition. *Proc Am Phil Soc, 67:*157, 1928.

Henderson, A. M. and Corruccini, R. S.: Relationship between tooth size and body size in American blacks. *J Dent Res, 55:*94, 1976.

Hiiemae, K. M. and Kay, R. F.: Evolutionary trends in the dynamics of primate mastication. *Symp Fourth Int Cong Primatology.* Basel, Karger, 1973, pp. 28-64.

Hine, K., Lavelle, C. L. B., and Flinn, R. M.: The analysis of tooth shape. *J Anat, 108:*585, 1970.

Holloway, R. L.: Tools and teeth, some speculations regarding canine reduction. *Am Anthropol, 69:*63, 1967.

Hooton, E. A.: *Up from the Ape.* New York, Macmillan, 1946.

———: *Up from the Ape.* New York, Macmillan, 1947, revised edition.

Howard, C.: A study of jaw and arch development considered with the normal and abnormal skeleton. *Int J Orthod, 12:*1, 1926.

Howard, R. D.: The unerupted incisor. *Dent Pract, 17:*332, 1967.

Howell, F. C.: In Howell, F. C. and Bourliere, F. (Eds.): *Transcript of Discussion of African Ecology and Human Evolution.* Chicago, Aldine, 1963, pp. 547-554.

Howell, F. C.: Review of "Man-apes or Ape-man." *Am J Phys Anthropol, 27:*95, 1967.

Hrdlicka, A.: Studies on tooth morphology. *Am J Phys Anthropol, 3:*429, 1920.

Hrdlicka, A.: Variations in the dimensions of lower molars in man and anthropoid apes. *Am J Phys Anthropol, 6:*423, 1923.

Hume, W. J.: Oligodontia. *Br. Dent J, 132*:71, 1972.

Hunt, E. E.: Malocclusion and civisation. *Am J Orthod, 47*:406, 1960.

Hurme, V. O.: Standards of variation in the eruption of the first six permanent teeth. *Child Dev, 19*:213, 1948.

Hurme, V. O. and Van Wagenen, G.: Basic data on the sequence of deciduous teeth in the monkey. *Proc Am Phil Soc, 97*:291, 1953.

Hutchison, A. C. W.: A case of total anodontia of the permanent dentition. *Br Dent J, 94*:16, 1953.

Illitis, H.: Inheritance of missing incisors. *J Hered, 39*:363, 1948.

Jolly, G. J.: The seed-eaters: a new model of hominid differentiation based on a baboon analogy. *Man, 5*:1, 1970.

Jorgensen, K. D.: The dryopithecus pattern in recent danes and dutchmen. *J Dent Res, 34*:195, 1955.

Kay, R. F. and Hiiemae, K.: Jaw movement and tooth use in recent and fossil primates. *Am J Phys Anthropol, 40*:227, 1974.

Keene, H. J.: Third molar agenesis, spacing and crowding of teeth and tooth size in caries-resistant naval recruits. *Am J Orthod, 50*:445, 1964.

Keene, H. J.: The relationship between third molar agenesis and the morphologic variability of the molar teeth. *Angle Orthod, 35*:289, 1965.

Keene, H. J.: Australopithecine dental dimensions in a contemporary population. *Am J Phys Anthropol, 27*:379, 1967.

Kenne, H. J.: The relationship between Carabelli's trait and the size, number and morphology of the maxillary molars. *Archs Oral Biol, 13*:1023, 1968.

Keith, A.: *The Antiquity of Man.* Philadelphia, Lippincott, 1925.

Kinzey, W. G.: Canine teeth of the monkey, *Callicabus moloch*: Lack of sexual dimorphism. *Primates, 13*:365, 1972.

Kinzey, W. G.: Ceboid models for the evolution of hominoid dentition. *J Human Evol, 3*:193, 1974.

Knott, V. B. and Meredith, H. V.: Statistics on eruption of the permanent dentition from serial data from North American white children. *Angle Orthod, 36*:68, 1966.

Kollar, E. J. and Baird, G. R.: The influence of the dental papilla on the development of tooth shape in embryo mouse tooth germs. *J Embryol Exp Morphol, 21*:131, 1969.

Kollar, E. J. and Baird, G. R.: Tissue interaction in developing mouse tooth germs. In Dahlberg, A. A. (Ed.): *Dental Morphology and Evolution.* Chicago, Chicago U Pr, 1971, pp. 15-29.

Korkhaus, G.: Die vererbung der kronenform und grosse menschlicher zahne. *Z Anat u Entwick, 91*:594, 1930.

Korenhof, C. A. W.: *Morphogenetical Aspects of the Upper Human Molar.* Utrecht, Uitgevers maatschappj Neerlandia, 1960.

Kovacs, Z.: A systematic description of dental roots. In Dahlberg, A. A. (Ed.): *Dental Morphology and Evolution.* Chicago, Chicago U Pr, 1971, pp. 211-256.

Kraus, B. S.: Occurrence of the Carabelli trait in south west ethnic groups. *Am J Phys Anthropol*, 17:117, 1959.

Kraus, B. S. and Furr, M. L.: Lower first premolars. *J Dent Res*, 32:554, 1953.

Kraus, B. S. and Jordan, R. E.: *The Human Dentition before Birth*. Philadelphia, Lea and Febiger, 1965.

Kraus, B. S., Jordan, R. E., and Abrams, L.: *Dental Anatomy and Occlusion*. Baltimore, Williams and Wilkins, 1969.

Lasker, G. W.: Genetic analysis of racial traits of the teeth. *Cold Spr Harb Symp quant Biol*, 15:191, 1951.

Lauer, C.: The relationship of tooth size to body size in a population of rhesus monkeys (*Macaca mulatta*). *Am J Phys Anthropol*, 43:333, 1975.

Lavelle, C. L. B.: Analysis of attrition of adult human molars. *J Dent Res*, 49:822, 1970.

Lavelle, C. L. B.: Mandibular molar tooth dimensions in different racial groups. *Bull Group Int Rech Sci Stomatol*, 4:273, 1971.

Lavelle, C. L. B.: Secural trends in different racial groups. *Angle Orthod*, 42:19, 1972.

Lavelle, C. L. B.: Relationship between tooth and skull size. *J Dent Res*, 53:19, 1974.

Lavelle, C. L. B. and Moore, W. J.: The incidence of agenesis and polygenesis in the primate dentition. *Am J Phys Anthropol*, 38:671, 1973.

Lavelle, C. L. B. and Plant, C. G.: Comparison between the right and left sides of the dental arch. *J Dent Res*, 48:971, 1969.

Lavelle, C. L. B., Ashton, E. H., and Flinn, R. M.: Cusp pattern, tooth size and third molar agenesis in the mandibular dentition. *Archs Oral Biol*, 15:227, 1970.

Leakey, L. S. B.: A new Lower Pleistocene fossil primate from Kenya. *Ann Mag Nat Hist*, 4:689, 1962.

Le Blanc, S. A. and Black, B.: A long term trend in tooth size in the Eastern Mediterranean. *Am J Phys Anthropol*, 41:417, 1974.

Leighton, B. G. and Orth, D.: Eruption of deciduous teeth. *Practitioner*, 200:836, 1968.

Leutenegger, W.: Metric variability of the post canine dentition in Colobus monkeys. *Am J Phys Anthropol*, 35:91, 1971.

Lewis, A. B. and Grainger, R. M.: Sex-linked inheritance of tooth size: a family study. *Archs Oral Biol*, 12:539, 1967.

Lewontin, R. C.: On the measurement of relative variability. *Systematic Zool*, 15:141, 1965.

Lombardi, A. V.: A factor analysis of morphogenetic fields in the human dentition. *Am J Phys Anthropol*, 42:99, 1975.

Ludwig, F. J.: The second mandibular premolar: morphologic variation and inheritance. *J Dent Res*, 36:263, 1957.

Lysell, L., Magnussen, V., and Thilander, B.: Time and order of eruption of the primary teeth in a longitudinal study. *Odont Revy*, 13:217, 1962.

Mahler, P. E.: Molar size sequence in the pongid dentition. *Am J Phys Anthropol, 41*:491, 1974.
Main, J. H. P. and Adams, D.: Experiments on tooth eruption in the rat incisor. *Archs Oral Biol, 11*:163, 1966.
Mandeville, C.: Congenital absence of permanent maxillary lateral incisor teeth: a preliminary investigation. *Ann Eugen Lond, 15*:1, 1949.
Mason, W. A.: Social organisation of the South American monkey, *Callicebus moloch. Tulane Studies in Zoology, 13*:23, 1966.
McDonald, R. E.: Anodontia in hereditary ectodermal dysplasia. *J Hered, 40*:95, 1949.
McGown, T. and Keith, A.: *The Stone Age of Mount Carmel*. Oxford, Claredon Press, 1939.
Meredith, H. V. and Hixon, E. H.: Frequency, size and bilateralism of Carabelli's tubercle. *J Dent Res, 33*:435, 1954.
Miles, A. E. W.: The dentition in the assessment of the individual age in skeletal material. In Brothwell, D. R. (Ed.): *Dental Anthropology*. New York, Pergamon 1963, pp. 191-210.
Miller, J., Hobsen, P., and Gaskell, T. J.: A serial study of the chronology of exfoliation of deciduous teeth and eruption of permanent teeth. *Archs Oral Biol, 10*:805, 1965.
Millhom, J. A. and Stafne, E. C.: Incidence of supernumerary and congenitally missing lateral incisor teeth in eighty one cases of hare lip and cleft palate. *Am J Orthod, 27*:599, 1941.
Mills, J. R. E.: Occlusion and malocclusion in primates. In Borthwell, D. R. (Ed.): *Dental Anthropology*. New York, Pergamon, 1963, pp. 29-52.
Montagu, M. F. A.: The significance of the variability of the upper lateral incisor in man. *Hum Biol, 12*:323, 1940.
Moore, W. J., Adams, L. M., and Lavelle, C. L. B.: Head posture in the Hominoidea. *J Zool Lond, 169*:409, 1973.
Moorrees, C. F. A.: *The Aleut Dentition*. Cambridge, Harvard U Pr, 1957.
Moorrees, C. F. A.: Normal variation in dental development determined with reference to tooth eruption states. *J Dent Res, 44*:161, 1965.
Morgan, G. A.: Supernumerary upper central incisors. *Dent Dig, 52*: 673, 1946.
Morris, D. H.: On the deflecting wrinkle and the *Dryopithecus* pattern in human mandibular molars. *Am J Phys Anthropol, 32*:97, 1970.
Moss, J. P. and Picton, D. C. A.: Mesial drift in adult monkeys (*Macaca iris*) when forces from the cheeks and tongue had been eliminated. *Archs Oral Biol, 15*:979, 1970.
Murphy, T.: Gradients of dentine exposure in human molar tooth attrition. *Am J Phys Anthropol, 17*:179, 1959.
Nanda, R. S.: Eruption of human teeth. *Am J Orthod, 46*:363, 1960.
Nanda, R. S., Khan, I., and Anand, R.: Age changes in the occlusal pattern of the deciduous dentition. *J Dent Res, 52*:221, 1973.

Ness, A. R.: Movement and forces in tooth eruption. *Adv Oral Biol, 1*:33, 1964.

Nicholls, J. I., Daly, C. H., and Kydd, W. L.: Root surface measurement using a digital computer, *J Dent Res, 53*:1338, 1974.

Nissen, H. E. and Rissen, A. H.: The deciduous dentition of chimpanzees. *Growth, 9*:265, 1945.

Niswander, J. D. and Sujaku, C.: Dental eruption, stature and weight of Hiroshima children. *J Dent Res, 39*:5959, 1960.

Oppenheimer, A. M.: Reply to criticisms. *Curr Anthropol, 7*:357, 1966.

Orlosky, F. J., Swindler, D. R., and McCoy-Beck, H. A.: Metric trends of the anterior teeth in African monkeys. *Hum Biol, 46*:647, 1974.

Osborne, J. W.: Some genetic problems in interpreting the evolution of the human dentition. *J Dent Res, 46*:945, 1967.

Osborne, J. W.: The evolution of dentition. *Am Sci, 61*:548, 1973.

Peak, B. L.: An unerupted supernumerary tooth in the lower third molar region. *N Z Dent J, 63*:55, 1967.

Pedersen, P.O.: The East Greenland dentition, numerical variation and anatomy. *Meddele:ser om Gronland, 142*:1, 1949.

Perreault, J. G., Demirjian, A., and Jenicek, M.: Emergence des dents permanentes chez les enfants canadien-francais. *J Canad Dent Assoc, 40*:306, 1974.

Perzigian, A. J.: Natural selection on the dentition of an Arikara population. *Am J Phys Anthropol, 42*:63, 1975.

Perzigian, A. J.: The dentition of the Indian Knoll skeletal population. *Am J Phys Anthropol, 44*:113, 1976.

Picton, D. C. A.: Tilting movement of teeth during biting. *Archs Oral Biol, 7*:151, 1962.

Picton, D. C. A.: Some implications of normal tooth mobility during mastication. *Archs Oral Biol, 9*:565, 1964.

Porter, R.H.Y.: Univariate versus multivariate differences in tooth size according to sex, *J Dent Res, 51*:716, 1972.

Potter, R. H. and Nance, W. E.: A twin study of dental dimension. *Am J Phys Anthropol, 44*:391-396, 1976.

Rajcinova, E.: Investigation of Carabelli's sign among the populations. *Stomatologija, 54*:292, 1972.

Randall, F. E.: The skeletal and dental development and variability of the gorilla. *Hum Biol, 15*:236, 16:23, 1943-1944.

Remane, A.: Eine Seltsame Gebissanomalie bei einem Stummeloffen zugleich ein Beitrag zur Frage der Selektions wirking bei der Gebiss differenzierung. *Zsaugetierk, 1*:114, 1926.

Remane, A.: Studieren uber die Phylogenie des Menschlichen Eckzahns. *Z Anat u Entw Gesch, 82*:391, 1927.

Robinson, J. T.: The dentition of the Australopithecines. *Transv Mus Mem*, No. 9, 1956.

Rutimeyer, L.: Beitrage zur Kenntwiss der fossilen Pferde und sur vergleichenden Odontographie der Hufthiere uberkaupt. *Verhnaturf Ges,* 3:558, 1863.

Schour, I.: The hypophysis and the teeth. *Angle Orthod,* 4:3, 1934.

Schultz, A. H.: Eruption and decay of the permanent teeth in primates. *Am J Phys Anthropol,* 19:489, 1935.

Schultz, A. H.: *Primatologia.* Basel Karger, 1956.

Schultz, A. H.: Cranial and dental variability in *Colobus* monkeys. *Proc Zool Soc* (Lond), *130:*79, 1958.

Schultz, A. H.: *The Life of Primates.* New York, Universe Books, 1969.

Schuman, E. L. and Brace, C. L.: Metric and morphologic variations in the dentition of the Liberian chimpanzee. *Hum Biol,* 26:239, 1954.

Scott, A. B.: Evolution of the premolar teeth in the Mammalia. *Proc Acad Nat Sci* (Philadelphia), 405, 1892.

Selmer-Olsen, A.: An odontological study of the Norwegian Lapps. *Shriftn Det Norske Videnshaps-Akademi Oslo,* 65:1, 1949.

Seipel, C. M.: Variation in tooth position. *Sven Tandlak Tidskr,* 39: suppl, 1946.

Shaw, J. C. M.: *The Teeth of the Bony Palate and the Mandible in the Bantu Races of South Africa.* London, John Bale, Sons and Danielsson, 1931.

Sirianni, J. E. and Swindler, D. R.: Inheritance of deciduous tooth size in *Macaca nemestrina. J Dent Res,* 52:179, 1973.

Sofaer, J. A., Chung, C. S., Niswander, J. D., and Runk, D. W.: Developmental interactions, size and agenesis among permanent maxillary incisors *Hum Biol,* 43:36, 1971.

Suarez, B. K. and Spence, M. A.: The genetics of hypodontia. *J Dent Res,* 53:781, 1974.

Suzuki, M. and Sakai, T.: On the protostylid of the Japanese. *Ziuruigaku Zassi,* 63:81, 1954.

Swindler, D. R., Gavan, J. A., and Turner, W. M.: Molar tooth size variability in African monkeys. *Hum Biol,* 35:104, 1963.

Swindler, D. R., McCoy, A., and Hornbeck, P. V.: The dentition of the baboon (*Papio anubis*). In Vagtborg, H. (Ed.): *The Baboon in Medical Research.* San Antonio, S. W. Foundation, 1967, pp. 133-150.

Swindler, D. R. and Orlosky, F. J.: Metric and morphological variability in the dentition of Colobine monkeys. *J Human Evol,* 3:135, 1974.

Tank, G. and Strovick, C. A.: Caries experience of children of one to six years old in two Oregon communities. *J Am Dent Assoc,* 69:749, 1964.

Thomas, N. R.: Chromophore markers in eruption studies. *J Dent Res,* 42: 1080, 1964.

Thomsen, S. O.: Missing teeth with special reference to the population of Tristan de Cunha. *Am J Phys Anthropol,* 10:155, 1952.

Vallois, H. V.: The Fontechevade fossil man. *Am J Phys Anthropol,* 7:339, 1950.

Van Reenen, J. F.: Dental features of a low caries primitive population. *J Dent Assoc S Afr,* 19:1, 1961.

Van Valen, L.: Intensities of selection in natural populations. *Proc Int Congr Genet,* 1:153, 1963.

Wakatsuki, E.: A morphological study by contour lines of the crown of deciduous Japanese teeth. *Shikwa Gaku,* 67:359, 1967.

Walker, A. C.: Locomotor Adaptations in Recent and Fossil Madagascan Lemurs. Doctor of Philosophy Thesis, University of London, 1967.

Watson, R. and Hardwick, C.: Hypodontia associated with cleft palate. *Br Dent J,* 130:77, 1971.

Weidenreich, F.: The dentition of *Sinanthropus pekinensis. Palaeont Sin,* D:1, 1937.

Weidenreich, F.: The Keilor skull: a Wadjak type from south east Australia. *Am J Phys Anthropol,* 3:21, 1945.

Williams, J. L.: New evidence of man's relationships to the anthropoid apes. *J Dent Res,* 8:289, 1928.

Wolpoff, M. H.: *Metric Trends in Hominid Dental Evolution.* Cleveland, Press of Case Western Reserve University, 1971.

Wortman, J. L.: Comparative anatomy of the teeth of the Veterbrata. *Am Syst Dent,* 1:351, 1886.

Wright, S.: On the roles of directed and random changes in gene frequencies in the genetics of populations. *Evolution,* 2:279, 1948.

Zanowiak, P. P.: Dynamics of dental occlusion in baboons. *J Dent Res,* 53: 1208, 1974.

Zingeser, M. R.: Odontometric characteristics of the howler monkey (*Alcuatta caraya*). *J Dent Res,* 46:975, 1967.

Zingeser, M. R.: Cercopithecoid canine tooth honing mechanism. *Am J Phys Anthropol,* 31:205, 1969.

Zuckerman, S.: Age changes in the chimpanzee, with special reference to growth of brain, eruption of teeth and estimation of age; with a note on the Taungs ape. *Proc Zool Soc Lond,* 1:1, 1928.

CHAPTER FIVE

THE CALCIFIED DENTAL TISSUES OF PRIMATES

INTRODUCTION

IN MAMMALS (Fig. 22) the main mass of a tooth is composed of a tissue known as *orthodentine* which is essentially a calcified collagen. Root dentine is overlain by *cementum*, another calcified collagen, into which are inserted the ends of collagen fiber bundles of the *periodontium*. The other ends of these fibre bundles are inserted into the bone of the *alveolus*. Thus, the periodontal space between bone and cementum is occupied by collagen fibers, blood vessels and other connective tissue elements, together constituting the periodontal ligament which functions as a shock-absorbing, suspensory mechanism for the tooth. The crown of the tooth is covered by *enamel*, which is very highly mineralized and is of a fundamentally different nature from the other calcified dental tissues.

Throughout the mammalian orders, the fundamental nature of each dental tissue remains the same. There are, nevertheless, certain minor variations in the histological features which have led to the suggestion of using such variations for taxonomic purposes. However, as yet no order has been scrutinized in sufficient detail for the proper evaluation of dental histology as a potential taxonomic tool. This is one of the purposes of the survey of primate tissue described below.

Because of the ravages of dental diseases, human dental tissues have been the subject of intensive research, and a very large amount of information on them is available. For this reason, human dental tissues are used here as a standard with which the same tissues of other primates may be compared. Existing literature on primate dental tissues is very limited and the descriptions given below are original. The conclusions from these observations will be considered with other literature in the final discussion.

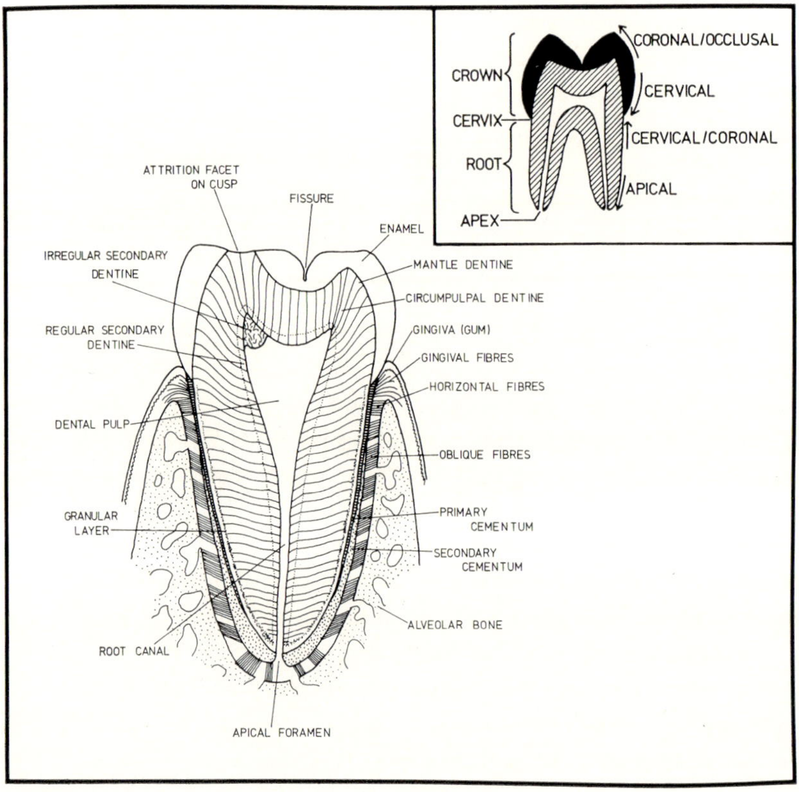

198 *Evolutionary Changes to the Primate Skull and Dentition*

Figure 22. Diagram of a human tooth *in situ* showing features of the enamel, dentine, cementum, and periodontium. Inset terminology used in this chapter.

THE DENTAL TISSUES OF MAN

Enamel

By weight, enamel consists of 95 percent mineral, 4 percent water, and 1 percent organic (mainly protein) material. By volume, these figures are 87 percent, 11 percent, and 2 percent respectively. Enamel is thus very highly mineralized and hard, the mineral being in the form of hydroxyapatite crystals which are easily visible in an electron microscope.

The basic structural unit is the *enamel prism,* or *rod* (Figs. 23 and 24) which runs from the amelodentinal junction to the surface of the enamel although, especially in deciduous teeth, the prism structure may disappear 15 to 20 μm from the surface in which case the outermost enamel coating is prismless and resembles the enamel of reptiles (Fig. 25). Each prism is approximately 5 μm in diameter and runs obliquely and occlusally from the amelodentinal junction (Fig. 26) to meet the enamel surface at an angle of approximately 60° in the midcrown region, tending towards 90° over the cusps and in the cervical regions. However, the course of a prism towards the surface is not direct but sinuous, in the transverse oblique plane. In the inner two thirds of the enamel, prisms are grouped in bands, each band being 10 to 20 prisms wide (Fig. 27). In adjacent bands, the sinuous prism curving occurs in opposite senses, such that in longitudinal sections of a tooth the prisms of one band are cut in a different plane from those in the two immediately adjacent bands. In this way, bands (Fig. 28) in which prisms are cut largely transversely (*diazones*) alternate with bands in which prisms are cut more longitudinally (*parazones*). These bands are well known as *Schreger* or *Hunter-Schreger* bands. Through the outer third of enamel the course of the prisms is more straight and parallel.

Two kinds of incremental markings may be recognized. *Cross-striations* occur at 3 to 5 μm intervals along each prism and represent daily growth increments (Figs. 2 and 3). Coarser markings (*striae of Retzius*) following a flattened, obligue, S-shaped course from the amelodentinal junction to the surface in longitudinal section, are spaced at 30 to 40 μm and occur at intervals of eight to ten days (Fig. 4). The striae are often brownish in color due to refractive effects and frequently, in deciduous teeth and the first permanent molar, one stria, the *neonatal line*, is especially noticeable (Fig. 27). Striae tend to be more prominent in outer enamel than elsewhere, where they curve towards the surface with a stepped or saw-tooth appearance. Each stria can be considered a "fossilized" record of the developing enamel front at one point in its history, the steps being related to the shapes of the formative ends (Tomes processes) of the ameloblasts. The steps become shallower and prisms less obvious as the surface is approached and the stria becomes smooth if it passes to the surface through prismless enamel (Fig. 25).

Occasionally, in longitudinal sections, it is found that the

Figure 23. *Homo.* Longitudinal ground section of human enamel prisms. Junctional lines between prisms and incremental cross-striations are regions differing, because of porosity, from the main prism mass in refractive index and so appear dark. Phase contrast: ×500.

whole enamel surface is covered with prismless enamel and, contrarily, occasionally it is completely absent, with prisms reaching the surface over the whole tooth. In the latter case, junctions between prisms near to the surface often seem to be enhanced compared with those in deeper enamel. Most commonly striae converge at the surface to give an appearance of overlapping "scales." The terminal wedge-shaped region of a "scale" is prism free while the intervening regions contain prisms. This arrangement is reflected in the surface topography in a series of incremental rings, around the crown, known as *perikymata.* Scanning electron microscopy reveals that the rings consist of smooth bands alternating with bands of depressions. The smooth areas are prismless and the depressions are the remains of pits in the

Figure 24. *Homo*. Enamel surface parallel with prisms; etched for 2 min with 0.1 N HCl. The junctional regions and cross striations have been preferentially etched, so that the prisms appear as beaded rods. SEM: ×550.

ends of prisms once occupied by the Tomes processes of the ameloblasts.

Prism structure may also be difficult to recognize in a thin layer of enamel, 15 to 20 μm thick, immediately adjacent to the amelodentinal junction (Fig. 26). The junction is usually markedly scalloped, with saucer-shaped depressions sinking into the dentine. Associated with the scalloping, but only seen well in transverse sections of the crown, are tuft-like regions of defectively mineralized enamel. Dentinal tubules terminate in branches in the mantle (outer) dentine. Some, however, cross the amelodentinal junction and penetrate into the enamel, for distances up to 10 to 15 μm, as the so-called *enamel spindles* (Figs. 26 and 27). The spacing between spindles is also 10 to 15 μm but varies from point to point in the same tooth.

202 *Evolutionary Changes to the Primate Skull and Dentition*

Figure 25. *Homo*. Enamel cut parallel with prisms, etched with N/70 HCl for several hours. Towards the natural enamel surface at the top of the picture, the prism structure disappears so that only incremental markings are seen. Both daily incremental markings (prism cross-striations) and longer, 8 to 10-day markings (striae of Retzius) are revealed by the etching. SEM: ×400.

All of the above features are visible with optical, bright-field microscopy. The hydroxyapatite crystals, however, are beyond the limits of resolution of optical microscopy. The crystals are readily observed with the electron microscope but because of the technical difficulties of sectioning such hard material, the information available on the ultrastructure of enamel is limited. Thus, the variability of prism outlines, crystal arrangement relative to prisms and crystal sizes has not yet been determined in any comprehensive fashion. Nevertheless, the principal features are now well known.

The stylized notion of the cross-sectional outline of an enamel

The Calcified Dental Tissues of Primates

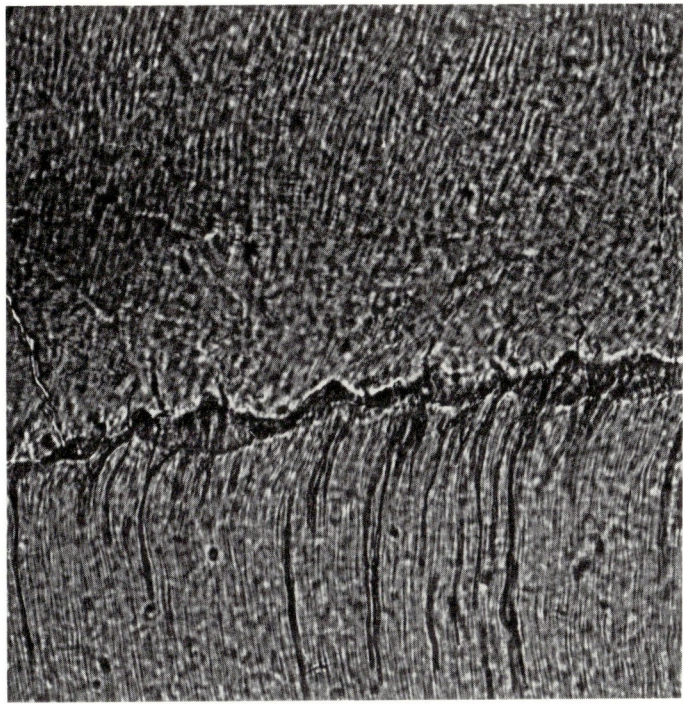

Figure 26. *Homo*. Longitudinal ground section of a tooth, showing the scalloped amelodentinal junction separating enamel (top), with prisms and cross-striations, from dentine (below). Dentinal tubules, some of which contain air, branch near the junction and a few cross into enamel as spindles. Sometimes it is difficult to identify prisms in the inner 10 to 15 μm of the enamel. Bright field: ×310.

prism is that of a keyhole or tadpole, with neighboring prisms interlocking (Figs. 29 and 31). Looking down on such cross-sectional outlines, the swollen head end lies towards the cuspal side of a prism, and the narrow tail towards the cervical side. This cross-sectional outline is common in human enamel but there are variations. Thus, the outline may be quite circular in prisms over a cusp, in which case there are areas of enamel, not contained within prism outlines, best described as interprismatic enamel. Again outlines may be only partial and they may be widely separated, so that it may be appropriate to apply the term interprismatic to any enamel which cannot be included within a

204 Evolutionary Changes to the Primate Skull and Dentition

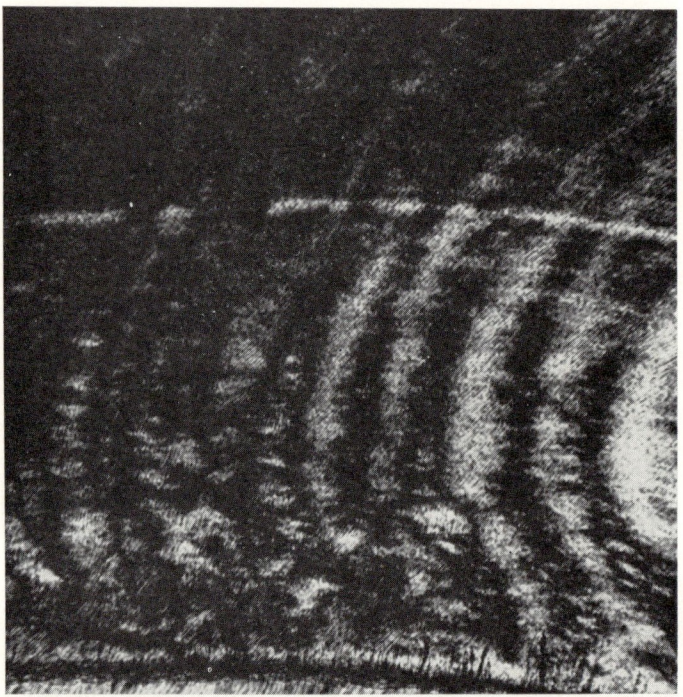

Figure 27. *Homo*. Longitudinal ground section of molar enamel viewed between cross polars. Alternating light and dark stripes reflect different prism directions in adjacent Hunter-Schreger bands (diazones and parazones). A neonatal line is seen in the enamel and at the amelodentinal junction are numerous spindles. Crossed polars: ×110.

complete, or perhaps easily extrapolated incomplete, prism boundary line. Other variations in prism outline, which may be crudely classified into four groups, are found among mammals (Boyde, 1965). These are shown in Figure 30 but it should be pointed out that, as in humans, more than one prism pattern may occur even in the enamel of one particular tooth.

The apatite crystals of human enamel are elongated hexagons of indeterminate length. In developing enamel, the ribbon-like crystals are known to be at least 20 to 30 μm long, but they increase in thickness with age to achieve cross-sectional dimensions of 100 × 150 nm. During sectioning, they fracture into segments

The Calcified Dental Tissues of Primates

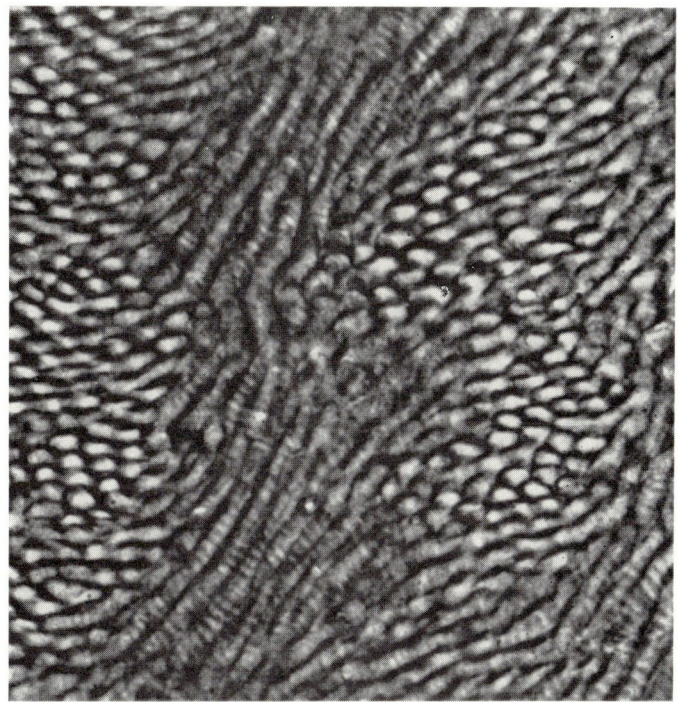

Figure 28. *Homo*. Longitudinal ground section of enamel demonstrating different prism orientations in the Hunter-Schreger bands; prisms are cut transversely in the diazones, longitudinally in the parazones. Phase contrast: ×800.

200 to 300 nm long, so that the true lengths in the fully formed tissue are unknown.

In the "head" of the prism, longitudinal crystal axes are parallel with prism axes but, on moving into the tail region, crystal direction slowly changes until crystal axes are at an angle of up to 60 to 70° from prism axes (Fig. 31). Functionally, this crystal deviation, together with prism decussation, will help to reduce a potential mechanical anisotropy which would be present in an enamel in which all prisms and crystals were straight and parallel.

Another consequence of the diverging crystal arrays in human enamel is that, from one prism to another, there is an abrupt

206 *Evolutionary Changes to the Primate Skull and Dentition*

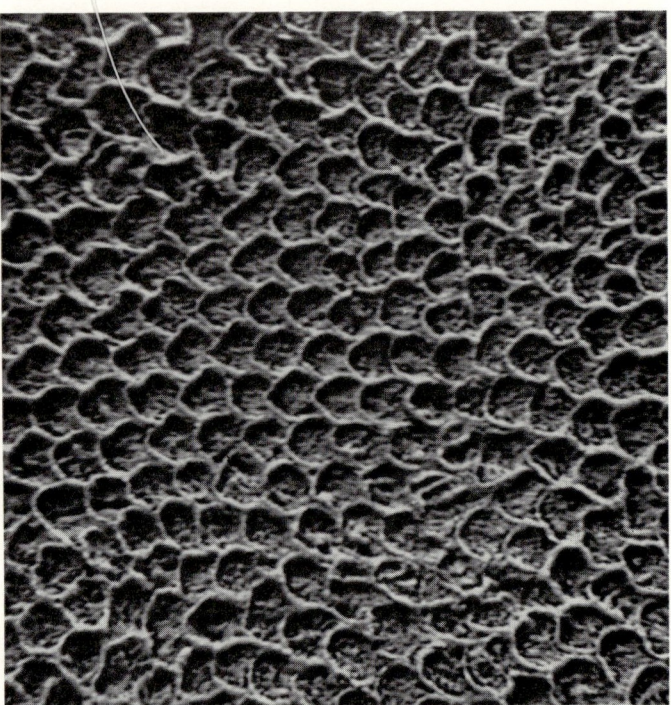

Figure 29. *Homo.* Acid etched enamel surface perpendicular to the prism direction showing the characteristic prism outlines. Those on the right alternate in position (pattern 3), whereas towards the center and left, they tend to be arranged in transverse rows (pattern 2). SEM: ×1,000.

change in crystal direction giving rise to discontinuities with which are associated water-filled pores. The refraction effect produced by these pores enables the prism junction to be seen optically. The optical appearance of the prism cross-striations and brown striae of Retzius are also due to variations in enamel porosity, although the disturbance causing a stria may sometimes be severe enough to cause an abrupt change in prism direction. Increasing porosity occurs both toward the inner half of the enamel and in the enamel near to the cervical margin.

The deviation in crystal orientation and differences in porosity are of particular value in comparative studies because both affect the double refraction seen in enamel. Hydroxyapatite is a uniaxial mineral with intrinsic negative birefringence. In simple enamels,

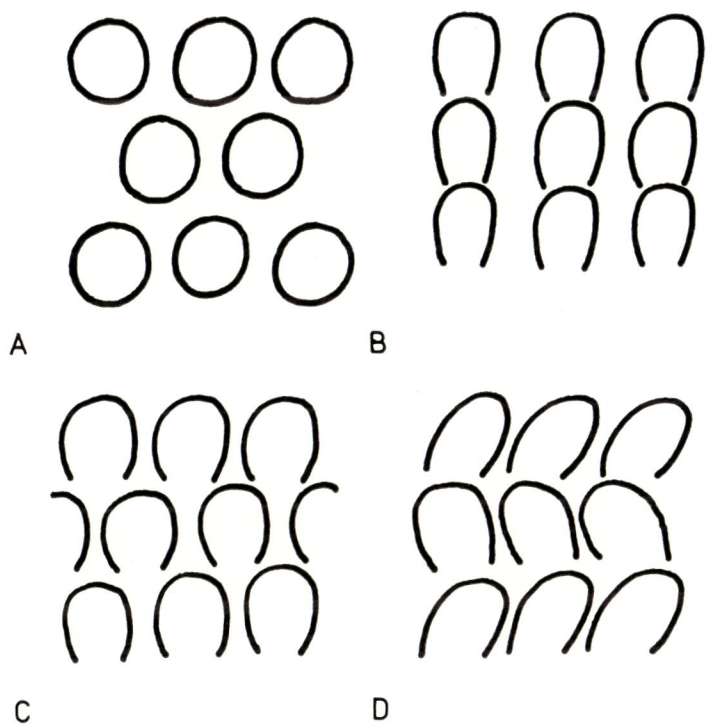

Figure 30. Patterns of prism cross-sections and arrangement found in mammals (after Boyde). (A) pattern 1 prisms, with circular outlines and wide interprismatic regions (found typically in Insectivora, Chiroptera, Sirenia, Odontoceti). (B) pattern 2 prisms, with incomplete boundaries, arranged in rows separated by interrow sheets of enamel (ungulates, marsupials). (C) pattern 3, with incomplete outlines, the prisms arranged in alternating fashion, so that interprismatic regions cannot be defined (man, elephants). (D) pattern 4 prisms, arranged in transverse rows, the incomplete cervical borders facing left and right in alternate rows (some rodents).

such as reptilian and the outer prismless enamel in many mammals, all the hydroxyapatite crystals are parallel with each other and stand perpendicular to the surface. Thus, within certain limits, between crossed polars this enamel is uniformly white, black, or grey according to whether the crystals are, respectively, at an angle to the polars of 45°, 0° or somewhere in between. Because of the divergence of crystals in human enamel, the prisms are white at 45° to the polars but striped black and white

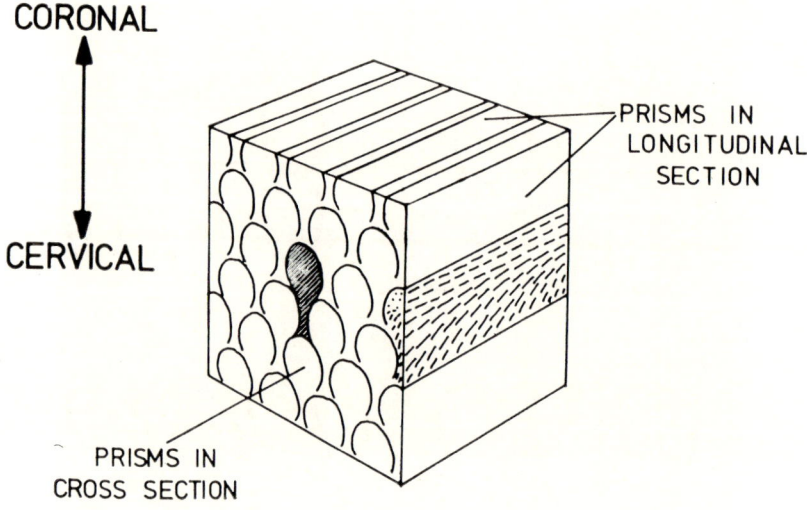

Figure 31. Diagram of a block of human enamel, to show the appearance of the prism junctions in different planes of section. The cross-hatched region indicates the "keyhole" appearance typically seen as the cross-sectional shape of the prisms. Elsewhere, the short lines represent crystal orientations, but the lengths of the lines are not proportional to crystal size (after Poole and Brooks; Meckel and Griebstein).

at or near the extinction position, with the black extinction stripes passing across the prisms as they are rotated past the 0° position. The striping seen in this way (Fig. 32) is a good indication of the presence of prisms as well as the arrangement of crystals within such prisms.

Because hydroxyapatite crystals are rod-shaped, they constitute a Wiener mixed-body system in which, if the refractive index of the rods is different from that of the substance separating them, a positive "form" birefringence is created. The strength of the form birefringence depends on the difference in refractive indices and the relative volumes of the two components. In well-mineralized enamel, the positive form birefringence is too small to interfere significantly with the intrinsic negative birefringence of hydroxyapatite. However, in inner enamel and cervical enamel, the porosity may be sufficiently large for the form birefringence to overcompensate the intrinsic birefringence to give an overall positive sign. For further details of these optical properties see Poole (1966).

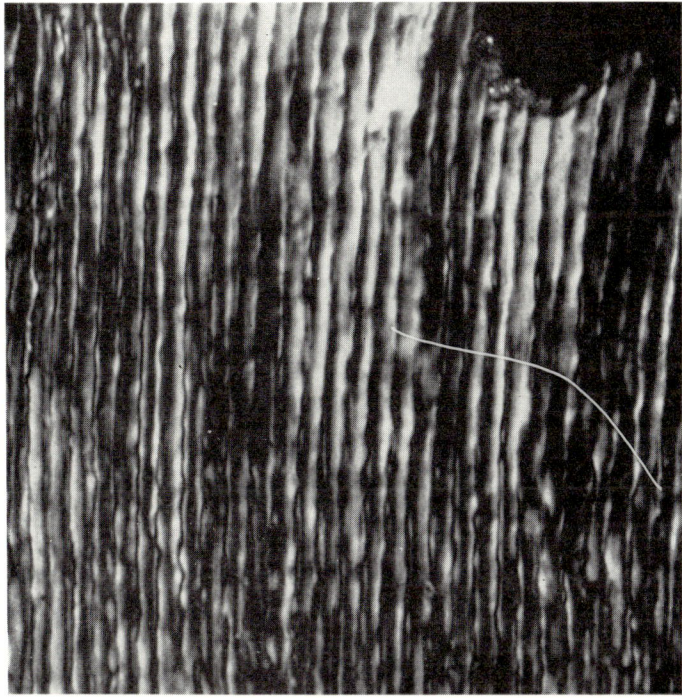

Figure 32. *Homo*. The same section as in Figure 2 but viewed between crossed polars. As explained in the text, the alternating black and white stripes are due to variations in the orientation of apatite crystals across the widths of the prisms. Crossed polars: ×500.

In this way it can be seen how the polarizing microscope can give a very rapid indication of crystal orientation and degree of mineralization in enamel. Likewise, it can be used to predict the ultrastructure of bone, cementum, and dentine, with the added complication that positive birefringence of collagen is very much greater than the negative birefringence of hydroxyapatite. Thus, although collagen amounts to only about 25 percent by volume in such tissues, and the mineral as much as 60 percent by volume, the overall birefringence is positive due to collagen.

Finally, the ultrastructure of enamel results in differential solubility. Thus, with brief exposure to concentrated acids and longer exposures to 5 to 10 percent EDTA at pH 7.4, the junctions of the prisms are preferentially dissolved, such that whole prisms may be chemically dissected in this way (Fig. 24). If a

210 *Evolutionary Changes to the Primate Skull and Dentition*

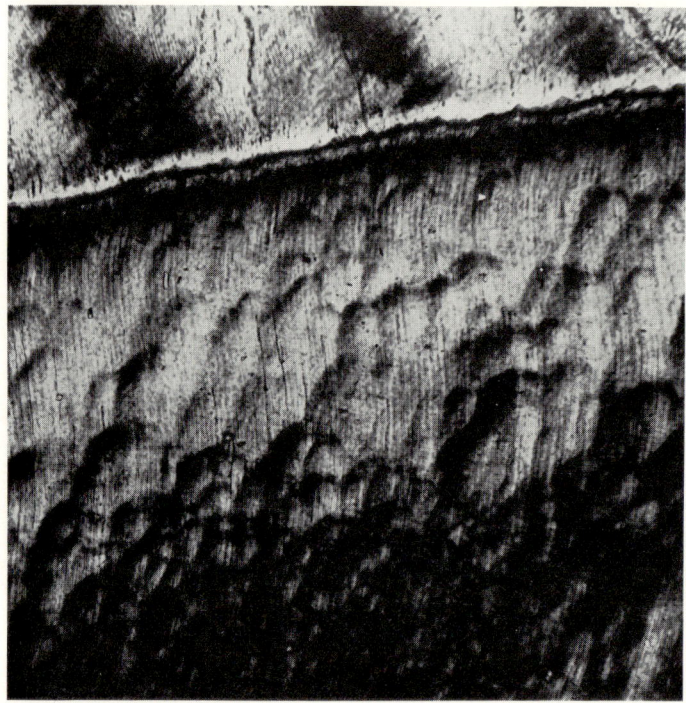

Figure 33. *Homo*. Longitudinal ground section of a molar, at the top showing prisms, Hunter-Schreger bands, and the inner bright layer of enamel adjacent to the amelodentinal junction. Below the junction is the narrow, grey zone of von Korff, separated by a black zone of compensation from the circumpulpal dentine making up the main mass of the tooth. In the circumpulpal dentine, tubules and globular calcospherites can be seen. Crossed polars: ×110.

surface is cut and polished perpendicularly to prism direction, the outlines of prism are made visible and they can be readily described by the use of scanning electron microscopy. Treatment with dilute acids tends to cause the centers of prisms to dissolve preferentially (Fig. 29).

Dentine

Dentine is the first tissue to form in any mammalian tooth germ and is laid down by *odontoblasts* which differentiate from the mesenchymal part of the tooth germ, the *dental papilla*. The

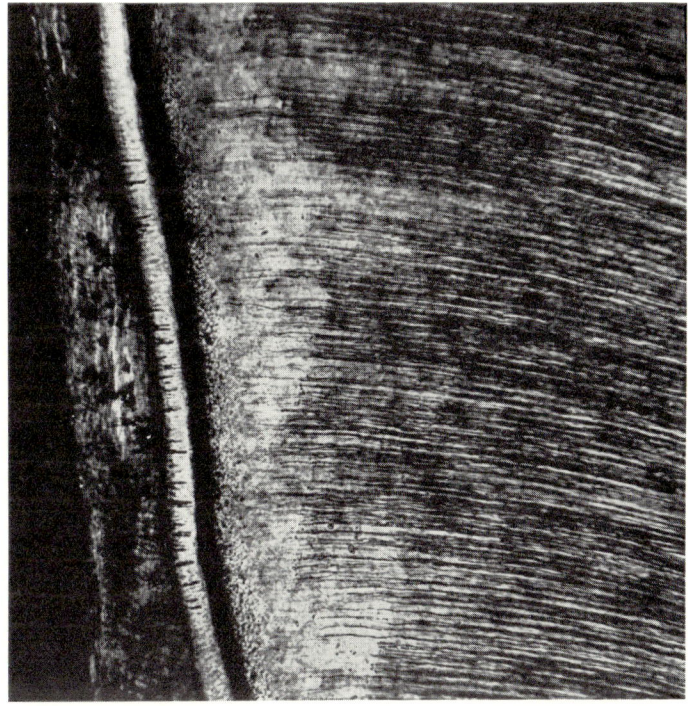

Figure 34. *Homo*. Longitudinal ground section of the root of a molar tooth, showing circumpulpal dentine. Passing towards the outside of the tooth (left), the layers encountered are: the granular layer of Tomes, the hyaline layer (black), primary cementum (white), and secondary cementum (grey). Crossed polars: ×110.

tissue thus grows in thickness *centripetally*, towards the papilla, in contrast to enamel which thickens *centrifugally*. Dentine formation proceeds by the deposition of a collagenous matrix, *predentine*, which subsequently mineralizes. The adontoblasts retreat in advance of the forming dentine but each cell leaves behind a process which occupies a *dentinal tubule* recognizable in the mature tissue (Fig. 34). The presence of tubules and the lack of included cells is a principal feature distinguishing dentine from bone and cementum. Odontoblasts have a limited functional life and, thus, dentinogenesis slows down in older regions and starts up in new regions. This means that the formative front of dentine does not lie parallel with either the outer surface of the

212 Evolutionary Changes to the Primate Skull and Dentition

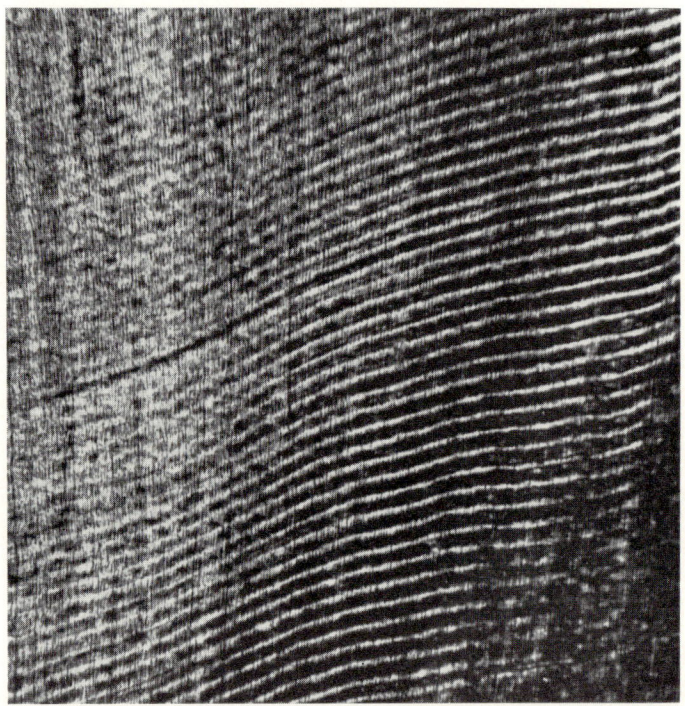

Figure 35. *Homo*. Longitudinal section through a molar tooth, showing incremental lines of von Ebner in the coronal circumpulpal dentine. Crossed polars: ×110.

tooth or with the surface of the mature pulp. The *incremental markings* which can be seen in dentine (see below) thus lie at an angle to these two surfaces.

In the crown the superficial, first-formed dentine, about 10 to 15 μm thick, is referred to as a *mantle* or *von Korff* dentine and is characterized by the presence of collagenous *von Korff* fibres which stand perpendicular to the amelodentinal junction (Fig. 33). During development, the von Korff fibers appear as coarse, argyrophilic fibres which are derived in part from the subodontoblastic layer of cells and are embedded in a network of fibers laid down by the odontoblasts. On the other hand, in the bulk of the dentine, known as *circumpulpal* or *von Ebner* dentine (Figs. 33 and 35), the matrix fibers are mainly orientated parallel with the forming front, at right angles to the dentinal tubules,

Figure 36. *Homo*. Fractured surface of dentine with longitudinally split dentinal tubules each surrounded by peritubular dentine smoother in texture than the irregularly fractured intertubular dentine. SEM: ×1,000.

although a few fibers lie parallel with the tubules. The behavior of dentine in ground sections under the polarizing microscope is dominated by the positive birefringence of the collagen so that the two layers of dentine are easily distinguishable with this instrument; the circumpulpal dentine is strongly positively birefringent relative to the incremental markings, while the mantle dentine is isotropic or slightly negative (Fig. 33). In many mammals, the mantle dentine contains prominent positively birefringent fiber bundles perpendicular to the amelodentinal junction.

The bulk of the dentine in the roots (Fig. 34), as in the crown, is made up of circumpulpal dentine. The superficial root dentine, however, has a different structure and is divisible into two layers. The outermost layer, known as the *hyaline layer of Hopewell-Smith* (Fig. 34), is about 10 μm thick in man. It possesses few tubules and appears homogeneous in ordinary light so is often difficult to distinguish from the adjoining layer of accellular cementum. In polarized light, however, the hyaline layer has an almost isotropic appearance like mantle dentine and can be recognized as a distinct layer. Underneath the hyaline layer lies the *granular Layer of Tomes* (Fig. 34), which appears to consist of dark granules in ground sections. This appearance is probably the effect produced by dilatations and contortions of the ends of the dentinal tubules.

In its mature state, dentine is a moderately elastic tissue somewhat harder than bone. It contains by volume 62 percent inorganic matter and 38 percent organic matter and water (in terms of weight, the values are 75 percent and 25 percent respectively). The inorganic phase consists of calcium phosphate in the form of apatite. The apatite crystals are plate-like, 3 nm thick and 20 to 100 nm wide, and therefore widely different in size and shape from those of enamel. The crystals are in intimate contact with the fibers of collagen, which make up about 80 percent of the organic phase of dentine. Mineralization of dentine occurs in two distinct patterns. In part, crystals are initiated and grow in relation to the matrix collagen fibers, so that the crystals and fibers have the same orientation in the mature tissue. However, a large proportion of dentine undergoes *globular* mineralization, in which apatite crystals grow radially from centers and thus cut across the direction of the collagen fibers. The *globules* or *calcospherites* are easily distinguishable in polarized light (Figs. 33 and 34). In the main body of the dentine, they are rarely spherical but have hyperboloid shapes, with the arcades directed outwards. The hyperboloid shape is presumably due to limitation of outward growth at an early stage by the inward advance of the mineralization front. In places, especially in the coronal dentine underneath the cusps, the dentine between groups of calcospherites may fail to mineralize; these regions are referred to as *interglobular dentine*. The occurrence of globular mineralization is a further feature distinguishing dentine from bone.

In ground sections viewed in polarized light, a series of regular bands can be observed in the dentine (Fig. 35). These bands have a narrow, fairly regular spacing and are referred to as *von Ebner* lines. Their appearance varies somewhat, however, with plane of section. They are prominent only in the inner two thirds of the coronal dentine. The lines represent rhythmical alterations in the organization of the matrix and are often termed incremental lines. Other incremental lines, which reflect variations in the mineralization process, and are thus independent of the von Ebner lines, can be detected in microradiographs. The dentinal tubules are not straight but follow an S-shaped curve, known as the *primary curvature*, between the inner and outer surfaces of the dentine. Superimposed on this curve are small undulations termed the *secondary curvatures*. In places, the secondary curvatures lie in register and this produces the optical effect of lines, known as *Owen's lines*, which cut across the direction of the tubules. The dentinal tubules give off small numbers of side branches within the circumpulpal dentine and branch more extensively at their outer ends in the mantle dentine (Fig. 26). Many of these terminal branches cross the amelodentinal junction and are continuous with enamel spindles.

Dentine is to be considered as a vital tissue since it is modified in response to aging, attrition and injury, as a result of the activity of the dental pulp tissue which it encloses. The bulk of the dentine, which is formed up to the time of completion of the roots (2-3 years after eruption) is referred to as *primary dentine* to distinguish it from *secondary dentine*, formed during later life. There are two forms of the latter tissue. *Regular* or *physiological* secondary dentine increases slowly in thickness with age, although its formation seems to be accelerated in areas underlying attrition. It contains about the same number of tubules as primary dentine and these are regularly arranged but they often lie at an angle to the primary dentinal tubules and this enables the junction between the primary and secondary layers to be detected. In *irregular* secondary dentine, on the other hand, tubules are fewer than in primary dentine, or even absent, and are twisted and irregular. Disturbance of the organization of the matrix is associated with the irregularity of the tubules and is revealed in polarized light by patchiness of the birefringence, in contrast to the uniform appearance of regular dentine. Irregular secondary dentine is laid down, generally speaking, over those parts of the pulp surface underlying areas of attrition or abrasion. It is also associated with caries of the enamel or primary dentine. The deposition of irregular secondary dentine, which seems to occur rapidly, may act to occlude tracts of tubules exposed by the above mentioned processes and thus protect the pulp from injury.

A process which begins before eruption, but which continues into old age, is the deposition of *peritubular dentine*, a hypermineralized layer formed on the walls of the dentinal tubules (Fig. 36). In man, the proportion of tubules lined with this tissue increases with age and particularly with attrition. However, some tubules may persist throughout life with no peritubular dentine. Some tubules become completely occluded with mineral with time but it is not known whether this is the result of continued formation of peritubular dentine or of some other process of mineral deposition. Tracts of dentine in which the tubules are completely occluded with mineral appear translucent when ground sections are examined in water. Such areas are particularly common in the roots and the frequency and extent of root translucency have been used in the estimation of age.

Tissues of Tooth Attachment

The attachment of mammalian teeth is effected by three tissues: comentum, which covers the roots of the teeth, the *periodontal ligament*, and *alveolar bone*. The periodontal ligament occupies the narrow space, about 150 to 200 μm wide, between tooth and bone and is made up of a system of fiber bundles, the *principal fibers*, penetrated by blood vessels and nerves. The principal fibers are inserted into alveolar bone at one end and into cementum at the other. The orientation of the fibers varies from place to place along the ligament. The portions of the principal fibers embedded in bone or cementum are known as *Sharpey fibers* and are partly or wholly mineralized.

Cementum formation begins in the region immediately coronal to the advancing edge of the root dentine and is carried out by *cementoblasts*, which are osteoblast-like cells derived from the tissues of the follicle surrounding the tooth germ. The cementoblasts lay down a collagenous matrix around the inner ends of the periodontal fibres and the tissue then mineralizes. The initial layer of cementum is *acellular* and is termed *primary cementum* (Fig. 34). The cementum reaches a greater thickness in the apical region of the root than in the coronal region and consists largely of *secondary* or *cellular cementum*, which is lacking in the coronal region. Secondary cementum contains *cementocytes* and thus resembles bone. Characteristically, the processes of the cementocytes are directed away from the tooth.

The Sharpey fibers present in cementum strongly influence the appearance of the tissue in polarized light. Since these fibers lie more or less perpendicular to the dentine surface, cementum tends to be negatively birefringent relative to this surface and is thus opposite in sign to the dentine. The birefringence of primary cementum is greater than that of secondary cementum (Fig. 34) because the latter tissue contains a higher proportion of matrix fibers, parallel with the root surface and at an angle of the Sharpey fibers. Both primary and secondary cementum show incremental lines.

The periodontal ligament and the alveolar bone are both in a condition of continuous turnover and remodelling. Indeed, the collagen of the ligament turns over much faster than in any other connective tissue of the body. By this turnover and remodelling, the position of the tooth is modified in response to attrition and growth and the teeth are thus maintained in their correct occlusal relationships. Cementum is not remodelled except in exceptional circumstances such as repair after injury. However, it is often observed that the orientation of the Sharpey fibers in cementum changes between successive incremental lines; this reflects the varying orientation of the principal fibers of the periodontal ligament in succeeding stages of the life history of the tooth.

Of necessity, this introduction to the structure of human calcified dental tissues is brief. A more complete general introduction is presented in Scott and Symons (1974). Details of the polarizing microscopy and ultrastructure of all tooth tissues are given in Miles (1967) and Schmidt and Keil (1971), while recent accounts of structure and development of enamel, dentine and attachment tissues are to be found, respectively, in Fearnhead and Stack (1971), Symons (1968) and Melcher and Bowen (1969).

THE DENTAL TISSUES OF NONHUMAN PRIMATES

Materials and Methods

About forty teeth from sixteen species were examined. Some of the material consisted of dried skulls or jaw bones, the rest of whole heads fixed in formalin. Some of the latter had been perfused initially in formalin-acetic acid and then stored in 70 percent ethanol. A list of the species studied, together with details of preservation and of the teeth studied, is presented in Table XVIII.

Most specimens were radiographed initially. The radiographs were useful in subsequent processing and allowed detailed assessment of the state of the dentition; this permitted the ages of the animals to be estimated in some cases. The dental tissues were studied by transmitted light microscopy, using both ordinary and polarized light, and by scanning electron microscopy (SEM). For light microscopy, ground sections were prepared and polished to a thickness of 50 to 80 μm using wet-and-dry carborundum paper and alumina powder. Prior to sectioning, many specimens were embedded in a supporting resin, usually cold-cure dental acrylic; this was removed, before the sections were mounted, by soaking in dimethylformamide or chloroform. Sections were prepared through teeth *in situ* from the fixed specimens and through single teeth from the dried specimens. Because the first permanent molars are present in the mouth for a greater part of the life span than any other tooth, these were the teeth of choice for sectioning; alterations in structure due to age and attrition could be studied in them. Sections of other teeth were also prepared where there was abundant material. Most sections were cut in a vertical labiolingual plane but sections in the mesiodistal and horizontal planes were cut from some teeth.

SEM was used to examine four types of preparation. Some teeth were split by compressing them in a vice and the resulting fracture surfaces, mainly in the longitudinal plane, were studied. The natural enamel surface was studied in certain species by examining unerupted teeth which had been removed from the jaws, soaked overnight in a 30 percent solution of sodium hypochlorite to remove organic material, and washed in water. Other teeth were split in half longitudinally, the cut faces were polished with alumina and etched with N HCl for 10 to 20 sec and then washed in water. Finally, on a tooth from each species, a flat surface in the enamel, parallel with the amelodentinal junction, was prepared by polishing and then etched as above. The fractured surfaces and etched surfaces were used to study the enamel prisms in the longitudinal and transverse planes. After their final wash in water, all specimens were soaked in 50 percent ethanol, allowed to dry, mounted on stubs and coated with gold/palladium by evaporation. They were examined in a scanning electron microscope (Cambridge S600) at 15 kV in the secondary emission mode.

TABLE XVIII

SPECIES OF PRIMATES STUDIED, WITH DETAILS OF PRESERVATION AND OF TEETH EMPLOYED

Family & Species	Preservation	Teeth
PROSIMII:		
Lemuridae		
Lemur sp.	Dried	First molar
Indriidae		
Propithecus sp.	Dried	First molar
Daubentoniidae		
Daubentonia madagascariensis	Dried	Lower incisor
Lorisidae		
Perodicticus potto	Dried	First molar
Galago senegalensis	Dried	First molar
ANTHROPOIDEA:		
Callithricidae		
Callithrix jacchus	Dried	First molar
Saguinus midas	Dried	First molar
Cebidae		
Saimiri sciureus	Formalin	1st, 2nd & 3rd molars
Ateles sp.	Dried	1st & 2nd molars
Cercopithecidae		
Cercopithecus pygerythrus	Formalin	First molar
Macaca fasciculata	Formalin	First molar
Macaca mulatta	Formalin	First molar
Macaca speciosa	Formalin	First molar
Papio cynocephalus	Formalin	1st & 3rd molars, 1st premolar, canine
Pongidae		
Gorilla gorilla	Dried	Third molar
Pan troglodytes	Formalin	First molar

Anthropoidea

Pongidae

In the *Gorilla* material examined, there was a very marked incremental line within the enamel, associated with a furrow at the tooth surface. Moreover, the enamel was not as hard as that of

Figure 37. *Gorilla*. Acid-etched surface perpendicular to the prism direction, with variation in etch pattern probably reflecting some minor variation in composition. The prisms are mostly alternating (pattern 3) but with some tendency for those on the left to lie in rows. SEM: ×225.

the chimpanzee. These observations suggest some form of enamel hypoplasia. However, the prism structure appeared normal and the dentine did not seem to be defective.

The enamel in both apes studied was made up of prisms 5 μm in diameter. In the body of the enamel, the prisms had a Type III morphology and were very regular, especially in *Gorilla* (Figs. 37 and 38) where the prism bodies were closer together than in *Pan*. At the enamel surface in *Pan*, but not in *Gorilla*, the prisms often had circular or spiral outlines (Fig. 39). Decussation of prisms was slight in *Gorilla* and Hunter-Schreger bands were not detectable by polarized light, although there was some poorly defined gnarling over the cusps. In the SEM, the prisms appeared almost straight and were largely parallel (Fig.

Figure 38. *Gorilla*. Part of the field of Figure 16, showing an almost perfect alternation of regular, closely packed tadpole-shaped prisms (pattern 3). SEM: ×2,250.

40). Hunter-Schreger bands, about ten prisms wide, were present in the inner two thirds of chimpanzee enamel except in the cervical half. In both species, the prisms in the outer enamel lay parallel and met the surface at right angles (Figs. 40, 41).

Retzius lines were well marked in both species (Fig. 41). Cross-striations were faint or absent in *Pan* but were very distinct in *Gorilla* (Fig. 39). In this species, there appeared to be doubling of the striations in many places, especially in the outer enamel. The interval between the cross-striations was about 5 μm in both species. In *Pan*, there was a pronounced incremental line which appeared to be a neonatal line since there was a corresponding mark in the dentine. In addition, there was a line near the cervical margin, possibly reflecting a disturbance of enamel formation during infancy.

222 *Evolutionary Changes to the Primate Skull and Dentition*

Figure 39. *Pan.* Acid-etched occlusal surface of molar. The prism outlines are circular or polygonal (pattern 1) and sometimes spiral. SEM: ×2,250.

The enamel of both species was marked by fairly extensive patches of positive birefringence in the inner regions and there was some positivity in the enamel cervical to the secondary disturbance line in the chimpanzee. A superficial layer of enamel showing an apparently prismless structure in ground sections was present in both species but was very thin (5-10 μm) and prism ends were revealed at the enamel surface in SEM specimens by light etching. The innermost 5 μm of the enamel showed a slightly enhanced birefringence and the border with the dentine was gently undulating. In *Pan*, spindles were present at intervals of about 10 μm over the whole crown and extended up to 40 μm into the enamel, whereas few spindles were observed in *Gorilla* except in the cuspal regions where they were generally about 15 μm long.

The dentine had a similar structure in both species. The

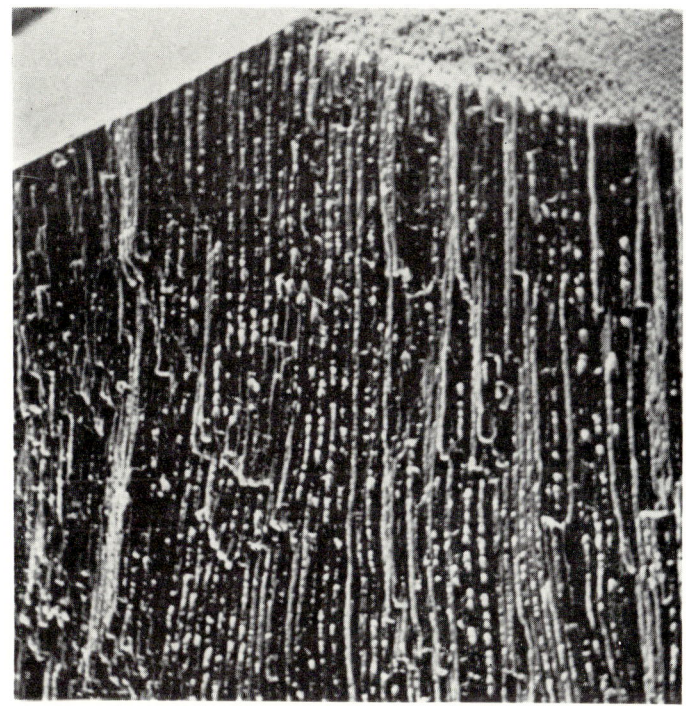

Figure 40. *Gorilla*. Acid-etched fractured surface of enamel with a longitudinal view of the parallel, almost straight prisms. Decussation of prisms does not occur and the prisms reach the natural enamel surface (left). SEM: ×225.

mantle dentine was 30 to 40 μm thick, had an opposite sign of birefringence to the circumpulpal dentine and appeared homogeneous in polarized light. The superficial dentine of the roots was made up of distinct granular and hyaline layers. Together these formed a zone 30 to 40 μm thick, of which the hyaline layer constituted about 15 μm (Figs. 44 and 45). Fiber bundles traversed the granular and hyaline layers, crossing coronally at about 45° to the root surface.

In the coronal dentine, the tubules were regular in size and structure and, except near the mantle dentine, they possessed almost no side branches. In root dentine, the tubules gave off numerous side branches and were extensively subdivided at their outer extremities. Many of the terminal branches mingled with

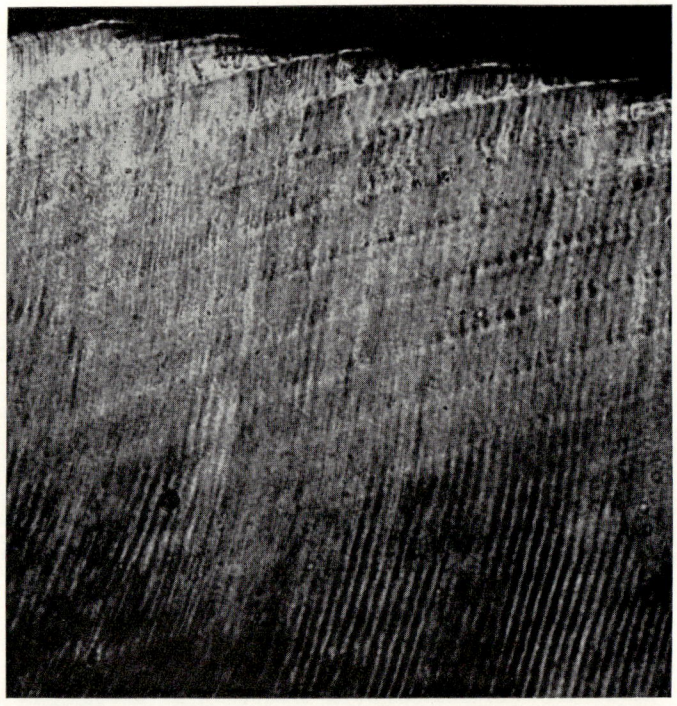

Figure 41. *Gorilla*. Outer part of enamel in ground section. Striae of Retzius form shallow perikymata on the enamel surface without a prismless layer. Deeper in the enamel, striping of the prisms indicates deviation in crystallite direction. Crossed polars: ×310.

the granular layer. In this region, the tubules were often twisted, looped or dilated (Fig. 45). This contributed to the granular appearance.

In polarized light, calcospherites were prominent in the circumpulpal dentine. At the outer surface, next to the junction with the mantle or granular layers (Fig. 48) these structures were small and formed a layer, the outer calcospherites being spherical, the inner ones hyperboloid. The outer third of the coronal dentine under this layer contained widely spaced, large hyperboloid calcospherites. This layer merged with the inner dentine, occupied by small, densely packed calcospherites tending to be spherical. In the roots, there was a zone free of calcospherites

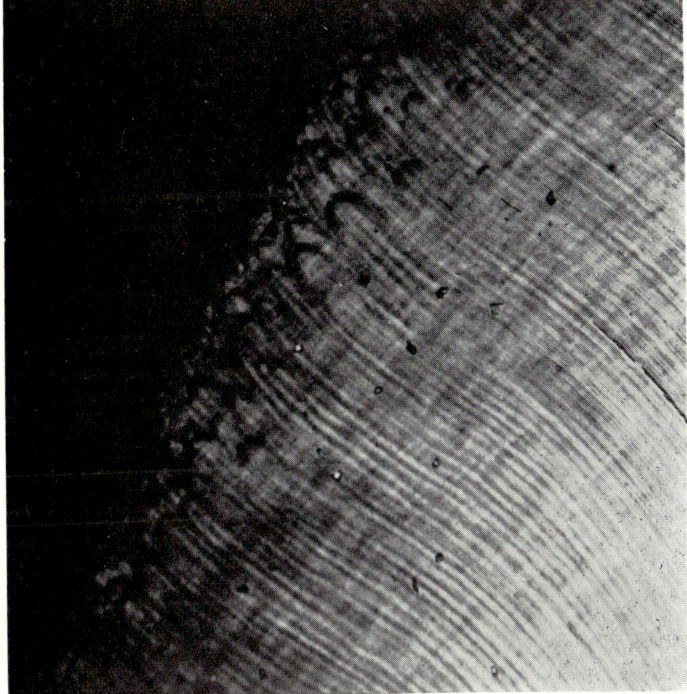

Figure 42. *Gorilla*. Coronal region of root in longitudinal section, showing circumpulpal dentine (light grey), containing a superficial layer of hyperboloid calcospherites. Note the dark and light striping of the dentine, the stripes running parallel with the tubules. The circumpulpal dentine gives way at the surface (left) to the hyaline layer (black) and a thin layer of primary cementum (faint grey). Crossed polars: ×110.

between the narrow outer layer and the thick, densely globular inner region. Interglobular dentine was observed only in one section of chimpanzee dentine.

A striking feature of the root dentine in *Gorilla*, observed nowhere else in this survey, was a pattern seen in the polarizing microscope of stripes running parallel with the dentine tubules (Fig. 42). These stripes appeared alternatively birefringent and isotropic in the diagonal position, but in the parallel position appeared alternatively negatively and positively birefringent; von Ebner lines were prominent in both species. Under the cusps,

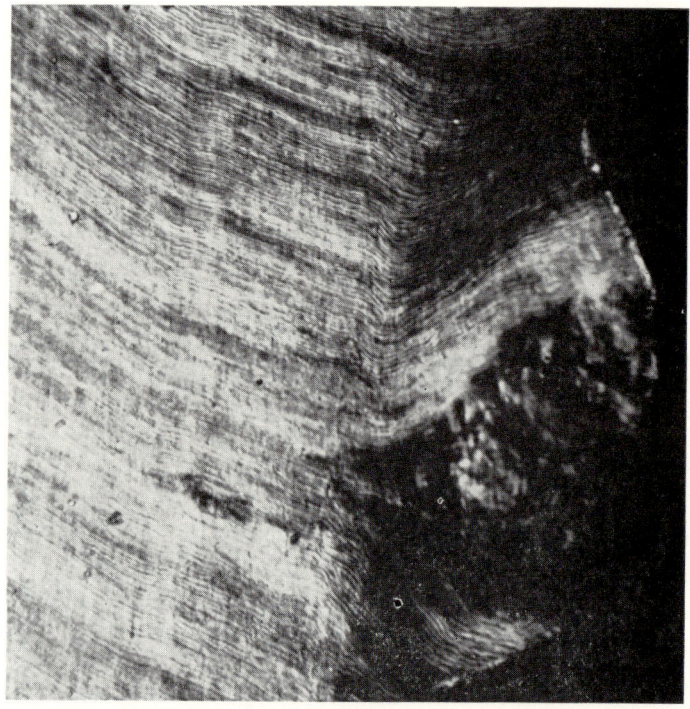

Figure 43. *Gorilla*. Longitudinal ground section through molar root. Primary dentine (left) gives way to regular secondary dentine (center) and then to irregular secondary dentine (right). The transition between each layer is marked by synchronous deviations of the dentinal tubules. Crossed polars: ×110.

these were sharply defined and were spaces at 5 μm intervals. Away from the cusps, they become more defined and broadened to about 20 μm thickness. Within the broadened lines, finer lines could be discerned at high power. In *Pan*, there was a marked neonatal line in the coronal dentine and in *Gorilla* a number of isotropic coarse lines spaced widely and irregularly.

Secondary dentine of both types was present in both species, although the tooth from *Gorilla* showed minimal signs of attrition. Regular secondary dentine formed a layer 80 to 100 μm thick in the coronal region, thinning off in the root canals. Irregular secondary dentine was most abundant on the floor of the pulp cavity and on the outer walls of the root canals (Fig. 43), reaching a

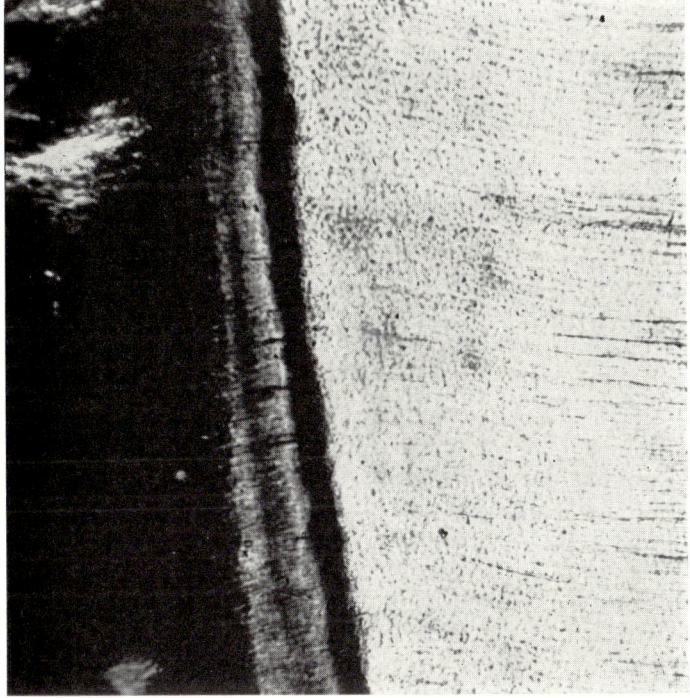

Figure 44. *Gorilla*. Longitudinal ground section of molar root. The primary cementum at this level, some way apical to the enamel, is becoming coarsely fibrous and is overlain by a thickening layer of acellular secondary cementum containing prominent Sharpey fibers embedded in an almost isotropic matrix. Crossed polars: ×110.

thickness of 400 to 500 μm. Several attached denticles were embedded in the irregular secondary dentine of the chimpanzee.

Primary cementum of typical structure was present in the coronal part of the root (Fig. 49). Apically, this merged with an extensive deposit of cementum reaching a thickness of over 100 μm (Fig. 44). It appeared that the fibers of the inner layer, instead of being closely packed as in the coronal region, became separated into discrete bundles, that is Sharpey fibers, which were embedded in a weakly birefringent matrix of opposite sign. The cementum external to the inner "primary" cementum had a similar structure and was irregularly stratified (Fig. 45). In *Pan*, the outer cementum was rich in cementocytes (Fig. 45) whereas

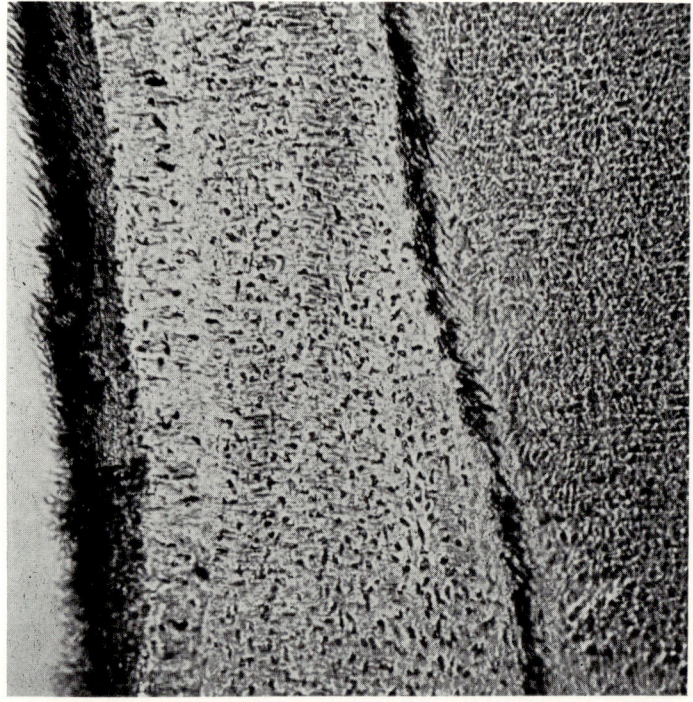

Figure 45. *Pan*. Longitudinal ground section through molar root, near apex. The dentine, with a prominent granular layer, is covered with a thick layer of richly cellular cementum. In the cementum, broad stratifications are visible as interruptions in the rows of Sharpey fibers. Note also the tubular constitution of the granular layer. Bright field: ×110.

cells were sparse in *Gorilla*. At the level of the alveolar crest, the periodontal ligament was 200 µm thick. The ligament was made up of principal fibers largely oriented obliquely.

Cercopithecidae

In this group the enamel was made up of Type III prisms, on average 5 µm in diameter (Fig. 46), although in *Papio*, for example, prism diameter varied widely, up to 10 µm. In *Macaca* there was a tendency for the prisms to be arranged in rows, giving a superficial impression of a Type II pattern, but interrow sheets of enamel were absent. There was decussation of prisms in all species. Hunter-Schreger bands (Fig. 47), about ten prisms wide, were prominent, especially at the sides of the teeth, and

The Calcified Dental Tissues of Primates

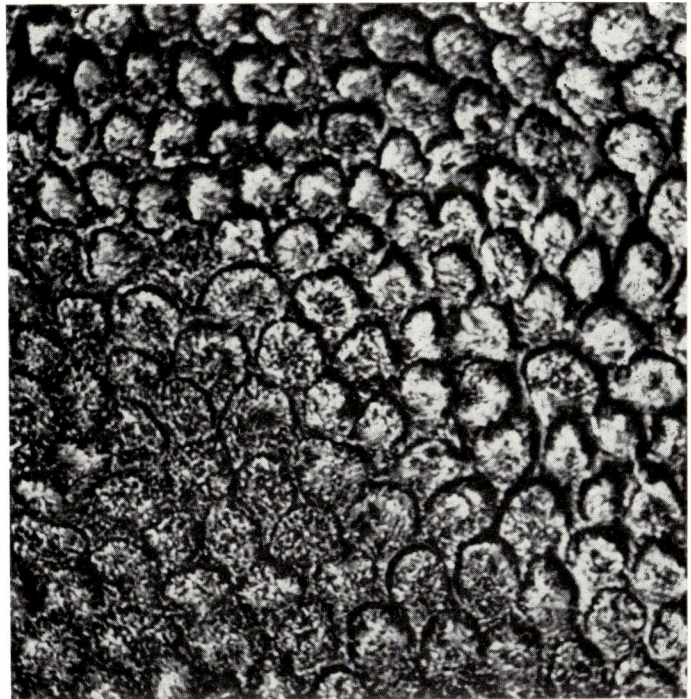

Figure 46. *Papio*. Acid-etched surface ground tangential to the enamel surface, with arcade-shaped prism outlines in alternating array (pattern 3). SEM: ×1,125.

occupied the inner half to two thirds of the enamel except in *Papio* and *Macaca fasciculata*. In these species, the bands extended about four fifths of the way to the surface. In the outer enamel, the prisms were parallel, inclined at 75 to 90° to the surface in *Cercopithecus* and *Macaca*, and 45 to 60° in *Papio*. The enamel in the region containing the Hunter-Schreger bands showed an overall reduction in birefringence and in some places a slight reversal of sign (Fig. 48). There were also occasionally areas of weak positive birefringence near the cervical margin in *Macaca*.

There was considerable variation in the incremental marking of the enamel. Cross-striations at 5 μm intervals were observed in all species (Fig. 48) but were particularly prominent in *Cercopithecus* and *M. fasciculata*. In these species, the striations often appeared to be subdivided into two or more fine lines. The

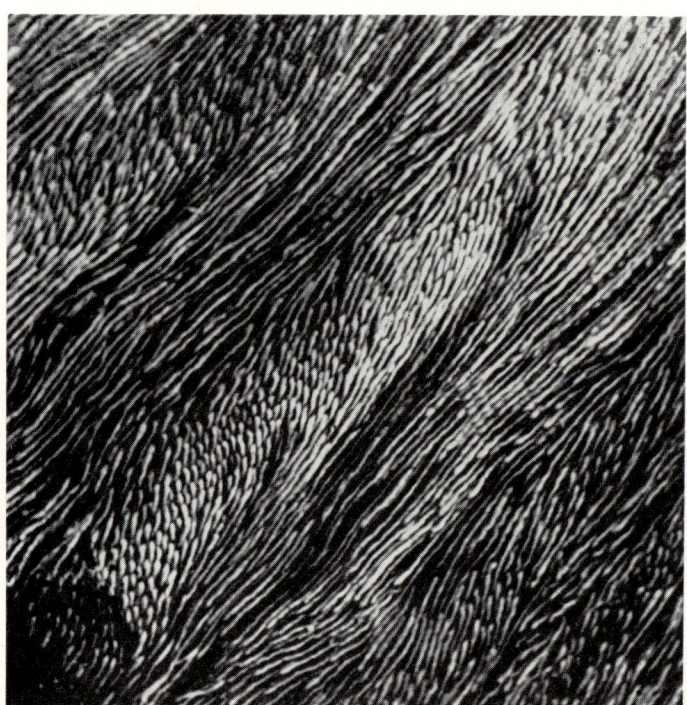

Figure 47. *Cercopithecus*. Ground and acid-etched longitudinal section through enamel, showing prisms arranged in Hunter-Schreger bands, each about ten prisms wide. Amelodentinal junction to bottom left. SEM: ×230.

striae of Retzius tended to be poorly defined. In *M. mulatta* and *M. speciosa*, the striae were generally discernible in the cervical enamel and at the surface, whereas only the superficial enamel contained striae in *M. fasciculata*, *Papio*, and *Cercopithecus*. The spacing of the striae varied from 15 μm in *M. mulatta* to 24 μm in *M. speciosa* and 30 to 32 μm in *Papio*.

In sections viewed in polarized light, a surface layer with distinctive properties was observed in all species, particularly in the enamel on the sides of the teeth (Fig. 48). The surface layer was 10 to 20 μ thick and was distinguished by having enhanced negative birefringence although in all species except *Cercopithecus*, there were frequent zones displaying positive birefringence (Fig. 48). Over the cervical half of the enamel surface, the superficial layer had a nonprismatic structure (Fig. 48). Penetration of the layer by the striae of Retzius frequently gave the layer a

Figure 48. *Macaca mulatta.* Longitudinal ground section of cervical enamel in which can be recognized a very broad band of prismless surface tissue, as well as prisms throughout the rest of the enamel, striae of Retzius and prism cross-striations. At the amelodentinal junction can be seen spindles running into the enamel from the almost black mantle dentine. Calcospherites and von Ebner lines are faintly visible in the circumpulpal dentine. In this region of the specimen, Hunter-Schreger bands cannot be seen. Crossed polars: ×110.

scaly or layered appearance (Fig. 48). Between the Retzius lines, there were often discernible finer striations about 2 μm apart. Towards the cusps, the superficial layer, while retaining its polarization properties, generally displayed a prismatic structure. In *Papio,* prismless regions were present here and there even near the cusps.

SEM examination of fractured specimens yielded more detail of the structure of the surface zone (Figs. 49 and 50). In *M. mulatta,* the surface zone in the cervical region appeared as a layer, 5 to 10 μm thick, made up of bundles of crystals standing perpendicular to the surface and diverging from the prism direc-

232 *Evolutionary Changes to the Primate Skull and Dentition*

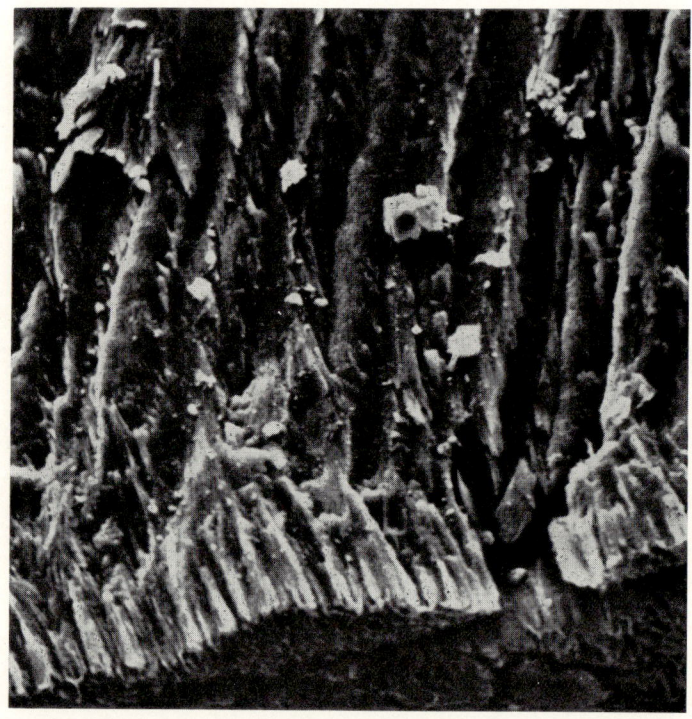

Figure 49. *Macaca mulatta*. Fractured outer part of enamel with prism structure ending towards the surface (below). The superficial enamel, 10 to 15 μm thick, has a prismless structure and consists of parallel crystals standing perpendicular to the surface and thus at an angle to the direction of the prisms. SEM: ×1,125.

tion (Fig. 49). The layer showed no trace of prism structure on the fractured surface, but on the enamel surface of cleaned unerupted teeth, there were circular depressions of similar diameter to the prism bodies.

In *Papio*, the enamel fractured obliquely along the striae of Retzius and it could be seen that, towards the surface, the layers of enamel between the striae lost their primatic structure and that the superficial zone was made up of overlapping layers of prismless enamel (Fig. 50). Clear signs of stratification within the layers were observed.

In all species, the amelodentinal junction was gently undulating rather than scalloped (Fig. 48), and was penetrated by

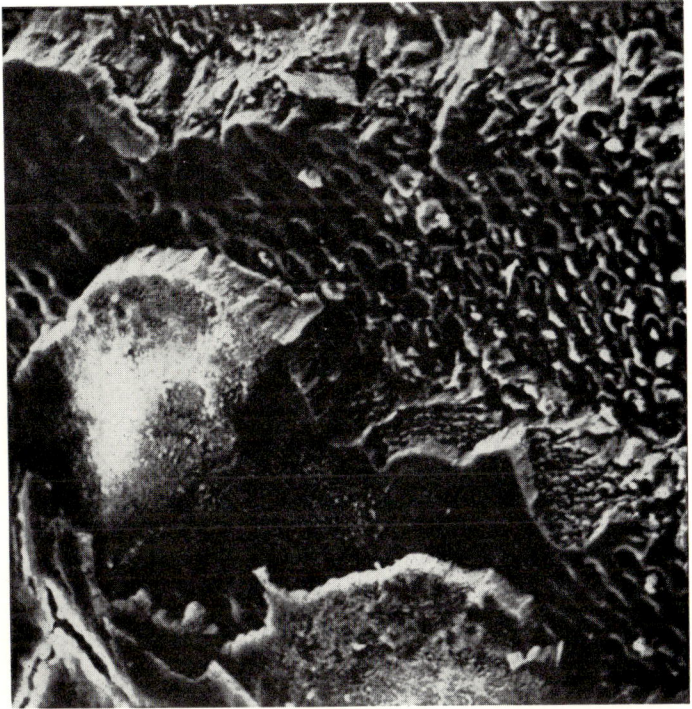

Figure 50. *Papio*. Fractured surface enamel, appearing stepped because of partial cleavage along successive striae of Retzius. In two of the cleavage planes, it will be observed that prism structure (right to left) disappears as the surface is approached. Note also some stratification of the prismless layer between successive striae. SEM: ×575.

numerous spindles, which extended 10 to 25 μm into the enamel and increased in frequency from the cervical margin to the cusps.

The mantle dentine was consistently 25 μm thick; the superficial layer at the roots, consisting of the granular and hyaline layers, was 30 to 32 μm thick (Fig. 53). In structure, these layers were very similar to their counterparts in the apes. This also applied to the structure of the circumpulpal dentine, although colcospherites were rather more numerous in the Old World monkeys. The von Ebner lines were prominent and had an appearance similar to those of the great apes.

The occurrence of secondary dentine was variable. In *M.*

234 *Evolutionary Changes to the Primate Skull and Dentition*

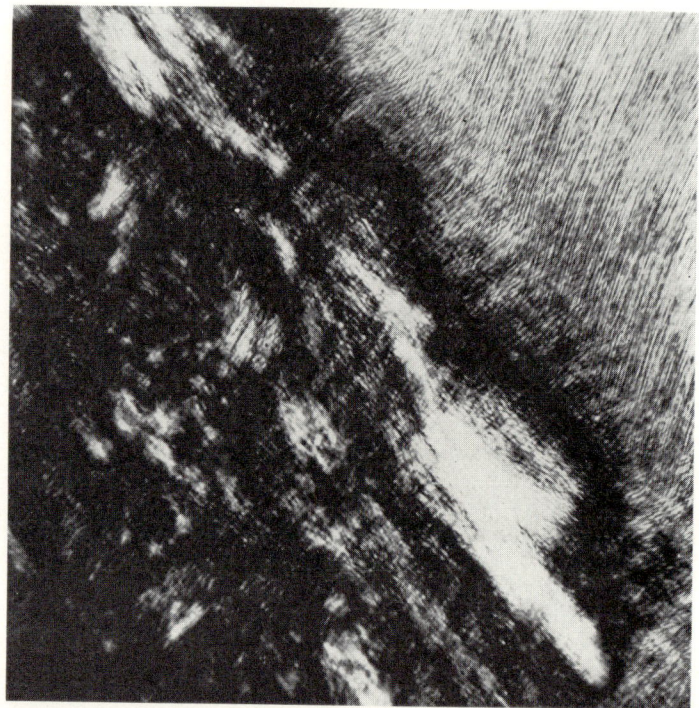

Figure 51. *Macaca fasciculata.* Longitudinal section of root dentine, showing regular (top right) and irregular (below left) secondary dentine, in which the irregular arrangement of the matrix fibers and crystals results in the patchy polarization image. Crossed polars: ×110.

speciosa, there was a distinct layer of regular secondary dentine 50 μm thick over the whole wall of the pulp chamber together with a mass of irregular secondary dentine reaching a thickness of 900 μm, underneath the attrition facet. A similar pattern was observed in the worn canines of one specimen of *Papio*, in which attrition had removed the whole of the primary dentine lingually; here the irregular secondary dentine formed a plug at the exposed pulpal surface. In the remaining specimens, regular secondary dentine was not distinguishable. Irregular secondary dentine was lacking in unworn or slightly worn teeth but in older teeth showing extensive attrition and exposure of primary dentine, irregular secondary dentine was present on the outer walls of the root canals, here reaching a thickness of 700 μm in some

Figure 52. *Papio*. Coronal dentine underneath a large attrition facet, fractured parallel with the tubules, each of which is patent but lined by a thin layer of smooth peritubular dentine. Intertubular dentine, comprising the mass of the dentine, has a highly irregular fracture surface. SEM: ×2,400.

specimens (Fig. 51), but was lacking from the walls of the pulp cavity.

In ground sections, the tubules underneath even large attrition facets were often seen, by the presence of air within them, to be patent, except for the outermost regions which seemed consistently to be occluded. This was confirmed by scanning electron microscopy of fractured dentine (Fig. 52). The SEM further revealed that the great majority of tubules were lined by a layer of peritubular dentine 0.5 to 1.2 μm thick (Fig. 52). Occasionally, tubules apparently filled with loose deposits of material were observed.

In recently erupted teeth, only primary cementum was found

236 *Evolutionary Changes to the Primate Skull and Dentition*

Figure 53. *Cercopithecus*. Longitudinal ground section through the root of a molar *in situ*. The circumpulpal dentine appears white (top right) and contains calcospherites. It is overlain by a hyaline layer, which appears black except for faint, diagonally oriented fibers, and a thick, stratified layer of cementum. Obliquely orientated principal fibers of the periodontal ligament are inserted into the cementum and into the alveolar bone which has a partly laminar and partly Haversian structure. Crossed polars: ×110.

on the root surfaces, generally forming a layer about 10 μm thick although it reached a thickness of 80 μm in *Papio*. In older teeth, the layer of primary cementum found coronally merged towards the apex with a thickening layer of stratified secondary cementum as in the Pongidae (Fig. 53). The total thickness of cementum near the apex was 100 to 250 μm. As in *Gorilla*, cementocytes were observed mainly in the cementum around the apex itself. Only in *Macaca fasciculata* were cementocytes at all prominent and in this species they occurred throughout the length of the tooth as well as at the apex.

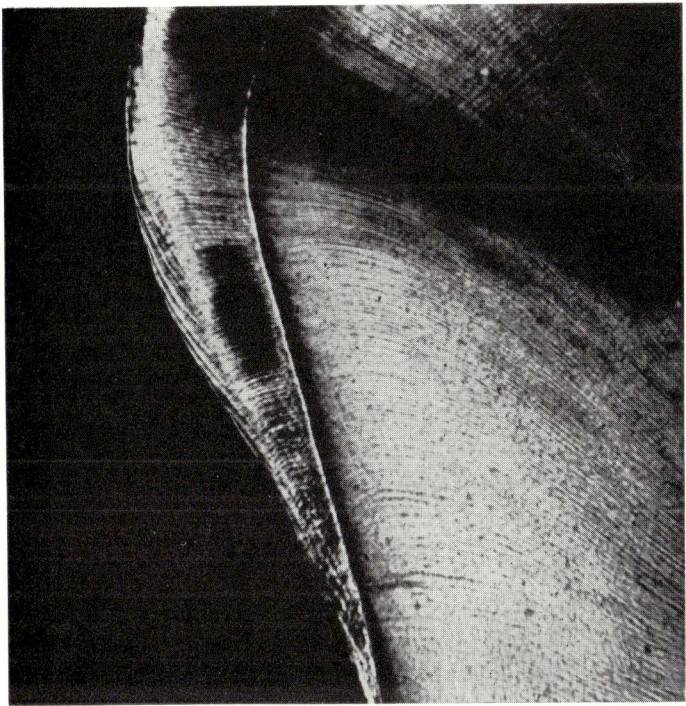

Figure 54. *Saimiri*. Longitudinal section of the cervical portion of the crown. In the enamel, the prisms run straight out from the amelodentinal junction without decussation, and enter at the surface a thick layer of prismless enamel. The black mantle dentine layer is pronounced, together with von Ebner lines in the circumpulpal dentine. Crossed polars: ×110.

The periodontal ligament was 200 to 250 μm wide at the alveolar margin in all species and was made up of prominent principal fibers uniformly obliquely orientated (Fig. 53). Often the fibers appeared distinctly wavy or crimped, especially in polarized light.

Cebidae

The enamel prisms in this group had a Type III morphology, the prism bodies having a diameter of about 5μm. The "heads" of the prisms were widely separated, giving an appearance of C-shaped prisms separated by interprismatic enamel rather than the keyhole appearance typical of the Old World monkeys and apes.

Often, the prism outlines were almost complete, the shape thus tending to be circular.

In *Ateles*, the inner half of the enamel was often isotropic with patches of positive birefringence. Within this layer were ill-defined Hunter-Schreger bands. The outer enamel was made up of parallel straight prisms, running at 75 to 80° to the surface. The enamel of *Saimiri* was much more regular (Fig. 54); over the whole tooth surface the prisms were parallel throughout the thickness of the enamel and stood almost perpendicular to the surface. Except for some areas which were largely isotropic, the enamel showed a uniform, negative birefringence.

Cross-striations at 5 μm intervals were present in both species. They were more pronounced in *Saimiri* and in this species, because of the highly ordered enamel structure the striations were aligned in rows running obliquely through the enamel (Fig. 54). Retzius lines were absent from both species, but lines at 8 to 12 μm intervals were present in the outer layer of the enamel, which had a scale-like appearance (Fig. 54). The outer layer of enamel was nonprismatic and, in fractured specimens of *Ateles* enamel, had a finely laminated structure (Fig. 55). It was interesting to observe on the enamel surface a shallow pitting suggestive of ameloblast depressions. In *Ateles*, the layer increased in thickness cervically, and the greater part of the enamel near the cervical margin, up to about 50 μm in thickness, lacked prisms. This region was penetrated by incremental lines which converged towards the cervical margin.

The mantle dentine was about 25 μm thick and the superficial root dentine 15 to 25 μm thick, both regions having the same structure as in the Old World monkeys and apes. The circumpulpal dentine contained numerous calcospherites, but they were relatively widely spaced and tended to be hyperboloid rather than spherical in form. Prominent von Ebner lines were present in both species (Fig. 54). In *Ateles*, they broadened to 12 to 16 μm apart between the cusps but in *Saimiri* they appeared as 5 μm thick lines, alternately dark and light, throughout their length. In mesiodistal sections of *Saimiri* molars, it was common to find, in the slopes of the cusps, a group of broad bands, 15 to 20 μm broad and 30 to 40 μm apart, showing enhanced birefringence (Fig. 56). These bands were associated with enhanced, coincident secondary curvatures of the dentinal tubules, and were thus taken to be Owen lines. The teeth available had suffered minimal attrition. A thin layer of regular secondary dentine was present in one tooth but the irregular variety was not observed at all.

Root surfaces were covered by a layer of cementum which was 10 to 15 μm thick on the outer surfaces of the roots (Fig. 57), but 25 to 50 μm on the inner surface. The cementum lacked cementocytes and largely resembled primary cementum, except in the thicker interradicular layers which contained some longitudinal fibers.

The periodontal ligament in *Saimiri* was made up of oblique fibers with some horizontal fibers occlusally (Fig. 57). In mesiodistal sections, the Sharpey fibers were present throughout the interdental ridges of bone (Fig. 57). The ligament was narrower than in the Cercopithecidae and Pongidae.

Callithricidae

There were a number of differences of enamel structure between the two species examined. The prisms were about 5 μm in diameter in both species, but size varied markedly in *Saguinus*. In *Callithrix*, the outlines of most prisms were circular, the cylindrical prism bodies being separated by interprismatic regions about 1 μm thick (Figs. 58 and 59). Prism form in *Saguinus* tended to be irregular, ranging from circular outlines to horseshoe shapes with side openings. As in *Callithrix*, the prism bodies were widely separated.

In both species, the prisms were largely straight and orientated at about 80° to the surface. There was some decussation of prisms in *Callithrix*, but the Hunter-Schreger bands were inindistinct. The enamel of *Saguinus* showed uniform negative birefringence but that of *Callithrix* contained extensive areas of opposite sign. In both species, the enamel contained regular incremental markings at intervals of 5 μm in *Callithrix;* 5 and 10 μm in *Saguinus*. These lines converged and terminated in a well-marked superficial layer of prismless enamel (Fig. 60). The boundary between enamel and dentine showed only gentle un-

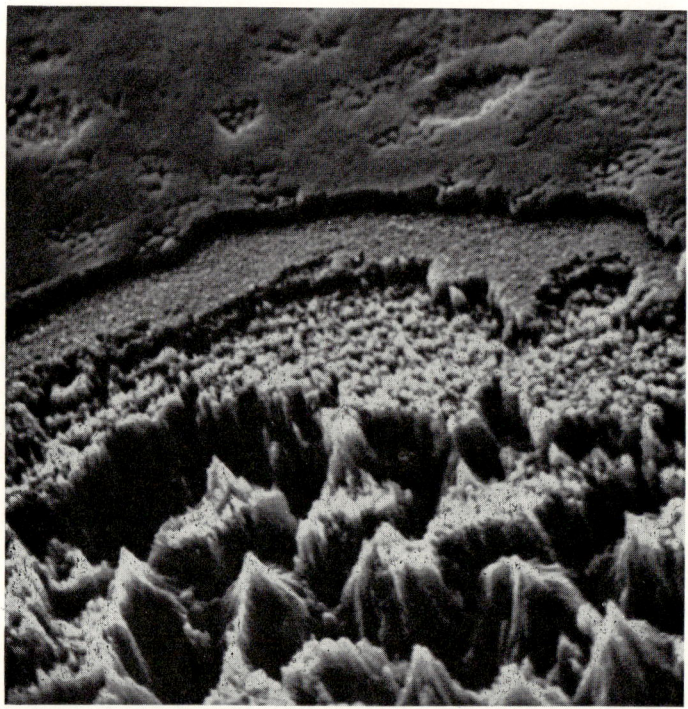

Figure 55. *Ateles*. Fractured surface of enamel of unerupted molar treated with hypochlorite. The surface is stepped because of partial cleavage along closely spaced striae. The superficial enamel is prismless and consists of closely packed crystals perpendicular to the surface. Shallow "stippled" imprints of cells survive on the surface. Thus, the existence of ameloblast depressions on the surface of enamel is not necessarily an indication that the prism structure extends right to the surface. SEM: ×2,000.

dulations and was penetrated in places by tubules from the dentine. These could be possibly regarded as enamel tubules rather than enamel spindles, because they were not dilated, but no tubules could be confidently identified beyond the innermost enamel.

An unexpected finding in the enamel of *Saguinus* was the identification by SEM of enamel lamellae (Fig. 61). These appeared as thin, sheet-like structures protruding from the etched surfaces ground parallel with the enamel surface. They ran parallel with the prisms throughout the thickness of the enamel and were associated with cracks on the natural enamel surface.

Figure 56. *Saimiri*. Coronal dentine, showing a thin, dark layer of mantle dentine covering circumpulpal dentine in which broad, striped bands indicating changes in matrix organization can be seen to correspond with synchronous displacements of dentinal tubules (Owen lines). Enamel, with prisms and cross-striations, is at the top of the field. Crossed polars: ×310.

The authors have not encountered lamellae in any other primate species.

The dentine in these species was almost identical. The mantle and hyaline layers were isotropic and 10 to 15 μm thick. A granular layer was lacking. These superficial layers contained small, narrow, hyperboloid calcospherites. In the circumpulpal dentine, calcospherites were confined to the coronal dentine, where they had the form of rounded arcades. In both species, there was a layer of regular secondary dentine, 20 to 30 μm thick, around the whole of the pulp cavity. Irregular secondary dentine was observed only in small amounts here and there, despite the fact that in the teeth sectioned, there was attrition of dentine.

A feature unique to *Saguinus* was the occurrence in the super-

Figure 57. *Saimiri*. Mesiodistal ground section through two adjacent molars *in situ*. Sharpey fibers from the periodontal ligament fibers extend deep into the interdental ridge of alveolar bone (patchy image, lower center) and give the impression of extending right through it as transalveolar fibers. On the left hand tooth, the periodontal fibers insert into dark grey primary cementum which overlies the black hyaline layer. In the outer circumpulpal dentine are elongated hyperholoid calcospherites. Crossed polars: ×110.

ficial dentine underneath the cervical enamel of fusiform dilations of the tubules (Fig. 62). Branched tubules arose from the outer ends of the dilated regions, many of the branches entering the cervical enamel. These structures were absent from the dentine of *Callithrix*. Owen lines, 20 to 30 μm apart, were present in the dentine near the cusps in both species. As in *Saimiri*, the lines consisted of exaggerated secondary curvatures lying in register. With the tubule curvatures were associated changes in the orientation of the matrix fibers which resulted in pronounced striping

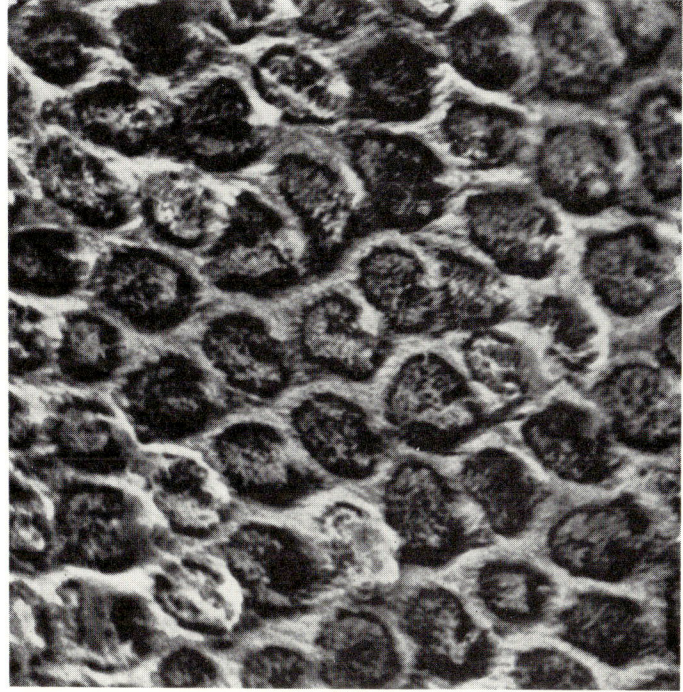

Figure 58. *Callithrix*. Acid-etched surface perpendicular to the prisms. The outlines are circular or polygonal (pattern 1), with wide regions of interprismatic enamel. SEM: ×2,300.

in polarized light. Von Ebner lines were somewhat diffuse but easily visible and had a periodicity of 5 to 8 µm.

The roots of the teeth were covered by a layer of cementum which was very thin in the cervical region but reached a thickness of about 30 µm towards the apices. Around the apices themselves, the cementum formed bulbous masses (Fig. 63). The cementum was thicker on the inner aspects of the roots than on the outer. The tissue was not divided into discrete primary and secondary layers. The cementum covering the greater part of the roots was acellular, showed no incremental markings and was only weakly birefringent, with a sign opposite to the dentine. The masses of cementum surrounding the apices were densely cellular and showed a patchy and irregular birefringence pattern.

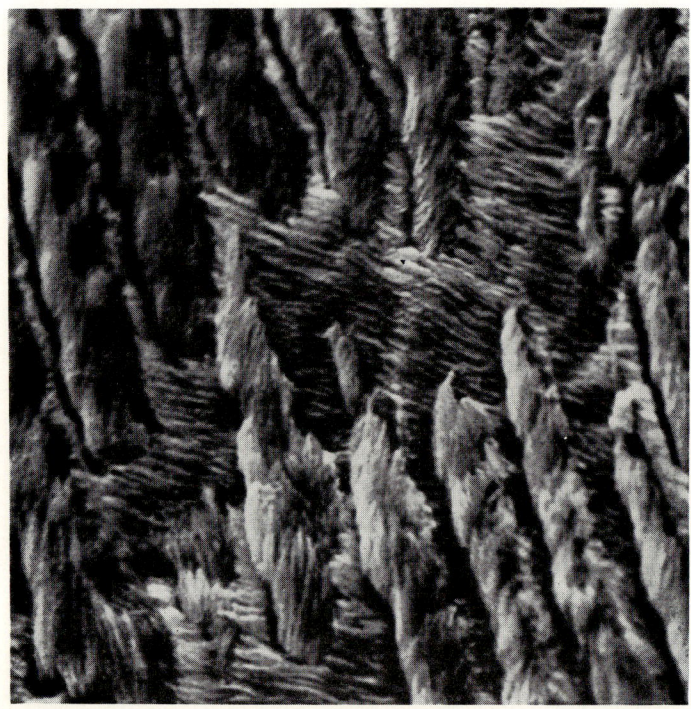

Figure 59. *Callithrix*. Acid-etched surface ground in longitudinal plane through enamel. Incremental markings may be seen as constrictions along the prisms. In this plane, the interprismatic enamel is seen as sheets of crystals which lie at an angle to the prism direction. SEM: ×2,300.

Prosimii

Enamel

The enamel of the lorisids *Galago* and *Perodicticus* was in many ways similar to that of the Cebidae and Callithricidae. The prisms, about 5 μm in diameter, generally had circular outlines, although some boundaries were incomplete as in the Type III pattern. There were wide interprismatic regions. In *Perodicticus*, there was prism decussation, which was, however, insufficient to produce distinct Hunter-Schreger bands (Fig. 64) and was more pronounced in the SEM than in the polarizing microscope; but in *Galago* the prisms were straight and parallel, running out at about 80° to the surface. Outside the region of prism decussation

Figure 60. *Saguinus*. Longitudinal ground section through lateral enamel, with straight, nondecussating prisms, closely-spaced incremental markings and a thick layer of prismless enamel. Elongated hyperboloid calcospherites are present in the outer dentine and faint Owen lines, due to kinking of tubules, are visible on the right. Crossed polars: ×310.

in *Perodicticus*, the prisms were straight and the enamel strongly negative except for the superficial enamel, which was positive due to form birefringence (Fig. 64). The superficial layer was nonprismatic in both species (Fig. 64). In the cervical region the enamel was negatively birefringent throughout its thickness. In both species, the middle region of the enamel contained extensive areas of weak or reversed birefringence (Fig. 64). Incremental markings were present at intervals of 5 μm in *Galago* and about 10 μm in *Perodicticus*. In the former species, they ran through the whole thickness of the enamel, whereas in the latter they were distinct only near the surface. In *Galago*, some of the striae were accentuated; this gave the enamel the appearance of

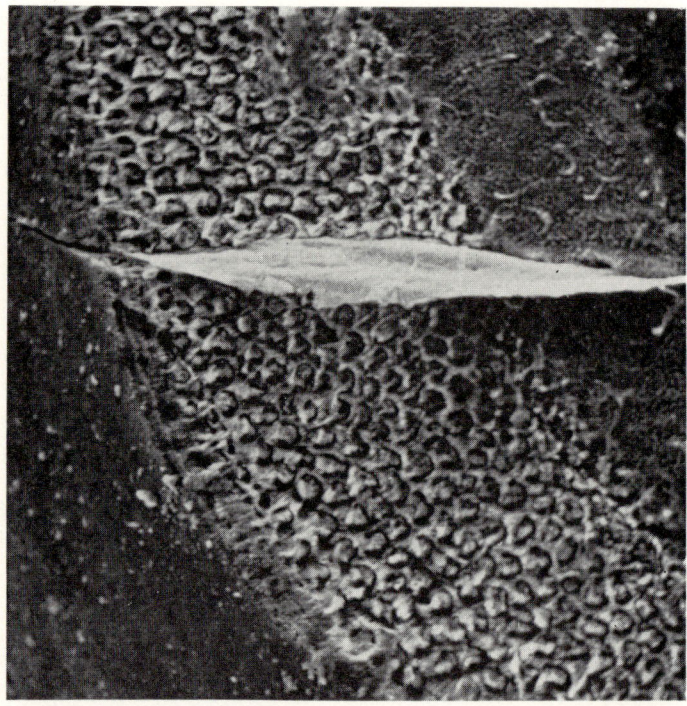

Figure 61. *Saguinus*. Acid-etched ground surface through enamel, revealing a lamella which appears as a sheet, standing proud of the etched surface and extending from the outer surface on the left, through the enamel parallel with the prisms, to the amelodentinal junction on the right. Some superficial prismless tissue overlies the prisms present throughout the rest of the enamel. SEM: ×55.

being made up of overlapping segments. The inner enamel was penetrated by tubules from the dentine which, in *Galago*, may have formed enamel tubules running well into the enamel but this point was difficult to verify because the prism boundaries were frequently accentuated and thus could be confused with tubules.

The enamel of *Propithecus* (Indriidae) resembled that of *Perodicticus* in most respects except for the reversal of birefringence in the superficial layer; this layer was also thinner and less well defined. Midway between consecutive 10 μm incremental markings, there occurred finer striations. The most interesting

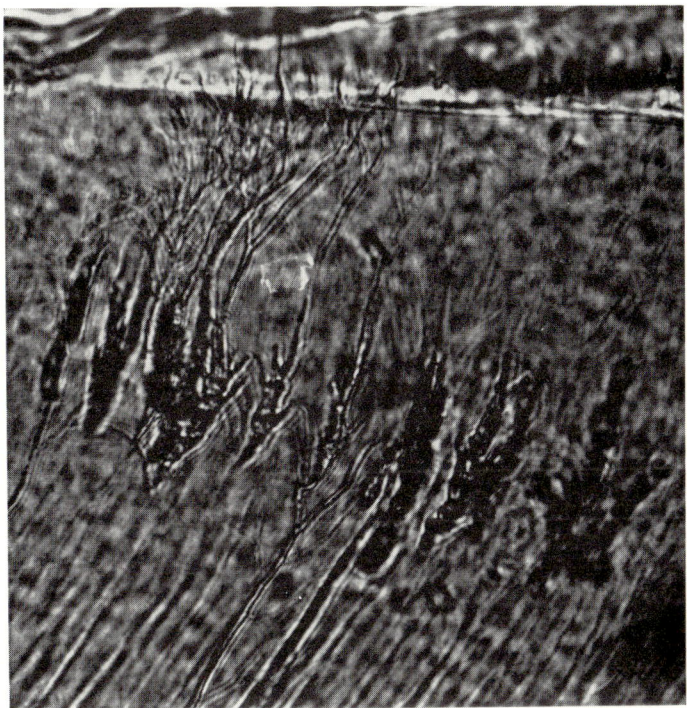

Figure 62. *Saguinus*. Ground section through the cervical region. Below the enamel are fusiform dilations of dentinal tubules. Branches from these pass through the mantle dentine and some cross the amelodentinal junction to enter the enamel. Bright field: ×700.

feature of the enamel in this species was the presence of numerous tubules running from the amelodontinal junction, where they were continuous with dentinal tubules, for a considerable distance into the enamel (Fig. 65). In the light microscope, the enamel tubules appeared fairly straight and parallel with the prism direction. They were most abundant at the occlusal surface and in the regions of the cusps. The tubules were filled with organic matter because in the SEM thin processes were observed lying exposed on the ground and etched surfaces of the tooth specimens (Figs. 66 and 67). Apart from establishing the presence of organic contents, the SEM confirmed that the tubules emerged from the dentine and showed that, in the cuspal region, the tubules were branched (Figs. 66 and 67). The branches often

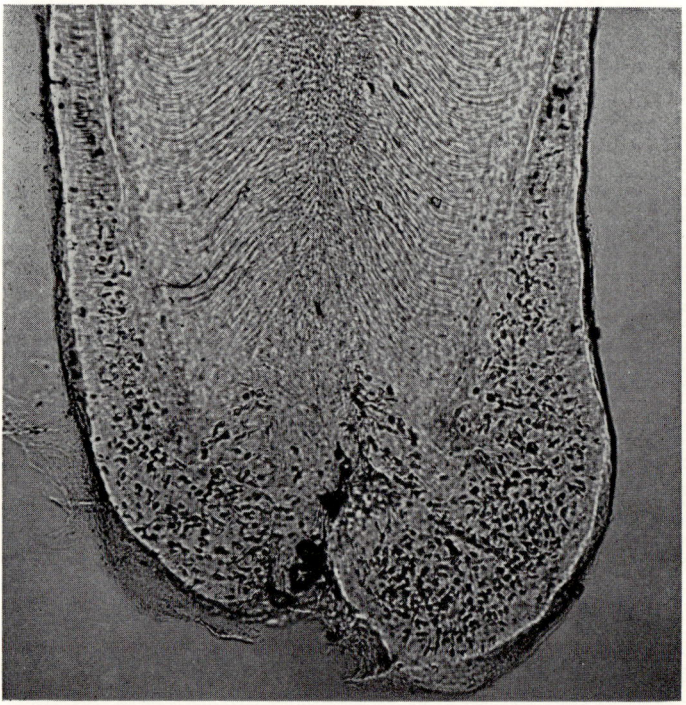

Figure 63. *Callithrix*. Longitudinal ground section of the root of a molar apex. The apex is surrounded by a thick mass of richly cellular cementum with Sharpey fibers visible in places. Coronally, the cementum thins off, more rapidly on the outer aspect (right) than on the inner, and becomes acellular. Bright field: ×110.

diverged from the prism direction and tapered off before reaching the external enamel surface, although fine processes were observed within 80 μm of this surface (Fig. 67).

In the specimen of *Lemur* examined, the enamel had an extremely unusual structure. In the polarizing microscope, the enamel appeared to be made up of parallel prisms running in gentle curves meeting the surface at 60 to 80° (Fig. 68). The prisms were intersected by marked incremental striae with a periodicity of about 5 μm (Fig. 68). There was a prismless surface zone which was well marked in places. On one side of the tooth studied, there was a prominent line, parallel with the incremental markings, associated with a dip in the surface (Fig. 68).

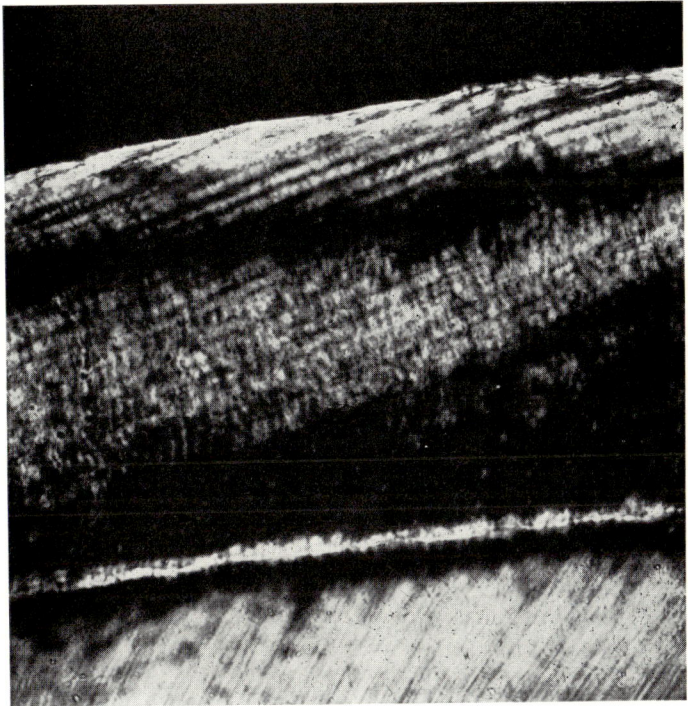

Figure 64. *Perodicticus*. Longitudinal ground section through lateral enamel. The prisms decussate to a slight degree without producing Hunter-Schreger bands visible in polarized light. Towards the surface, a dark compensation line separates the negative inner enamel from the outer positive enamel. In the superficial enamel, the incremental markings are pronounced and the outermost zone is prismless. Crossed polars: ×280.

The enamel immediately inside and occlusal to the line was positively birefringent. The area of this line was examined in the SEM on the prepared surface of the other half of the tooth from which the ground section had been prepared (Figs. 69 to 72). On a ground and etched surface tangential to the enamel, prism cross-sections with a very regular Type III morphology were observed on the occlusal side of the line. In the longitudinal plane, prism boundaries were largely obscure in this region, even after etching (Figs. 70 and 71). Instead, the action of the acid was to dissolve the enamel preferentially along numerous parallel striae running outwards and occlusally. The striae were spaced fairly

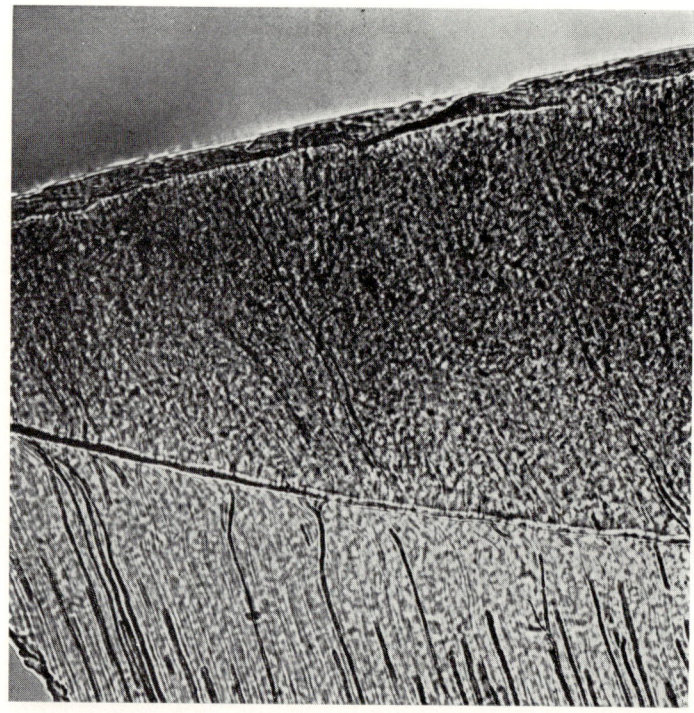

Figure 65. *Propithecus*. Longitudinal ground section through occlusal enamel. Tubules can be seen within the enamel and in a number of places show continuity with dentinal tubules. Bright field: ×280.

regularly at about 5 μm and thus corresponded to the incremental lines seen under polarized light. The striae were in places almost straight but often had an undulating or scalloped course. On the outer (cervical) aspect of the line, the enamel appeared to be nonprismatic and was again etched along the incremental striae (Fig. 72). In this region, the periodicity of the striae was less regular, although it was on average about 5 μm, and some of the inner striae converged cervically. The outer striae were straighter than the inner ones.

The enamel was penetrated by tubules from the dentine and some appeared to extend some way towards the exterior but, as with *Galago*, their identification was hampered by the occurrence of accentuated prism boundaries. No tubules were identified in the SEM.

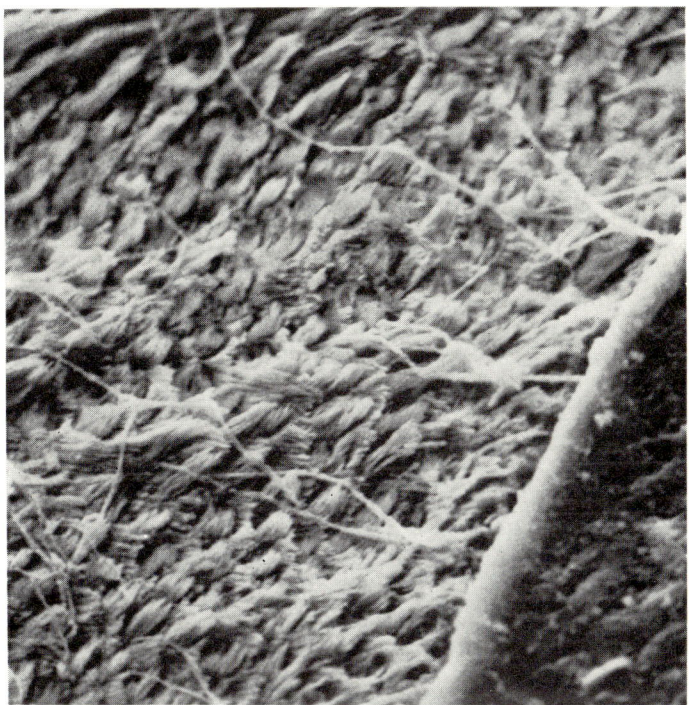

Figure 66. *Propithecus*. Acid-etched longitudinal ground surface through occlusal enamel near the cusp. The contents of enamel tubules, appearing as filamentous structures, pass radially into the enamel from the dentine (lower right) and branch. The orientation of these branches is unrelated to the directions of the prisms. Note the evidence of prism decussation in the varying plane of section of the prisms. SEM: ×1,150.

The *Daubentonia* material consisted of an incisor from which two transverse slices were cut; one was polished and mounted as a ground section, the other was prepared for examination by SEM. The enamel was distinguished by its elaborate, highly organized structure. The prisms appeared in cross-section as elliptical bodies arranged in rows and separated by wide sheets of interprismatic enamel (Fig. 73 and 75). The prism boundaries appeared to be complete but this may have been artifactual because the boundaries were etched unusually deep by our standard procedure. The prisms were organized into extremely regular Hunter-Schreger bands, which appeared in ground section as

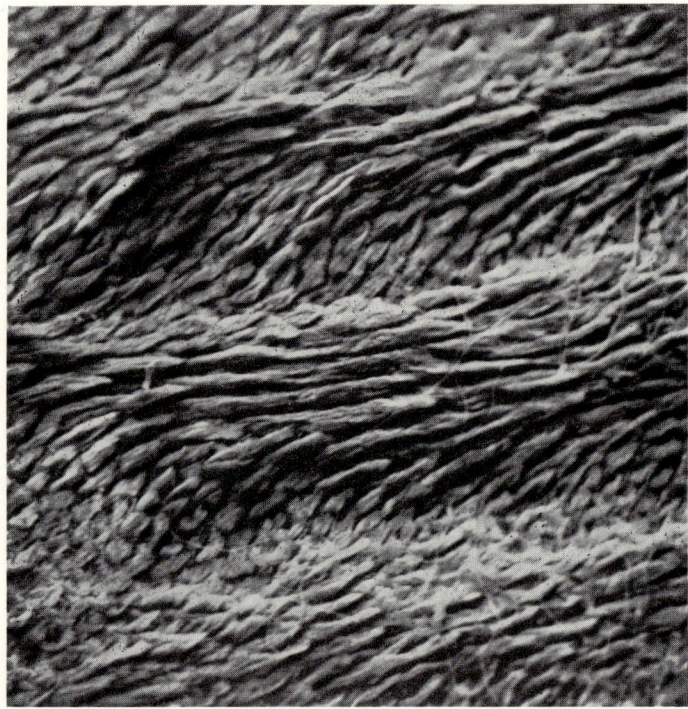

Figure 67. *Propithecus.* Outer enamel, from same specimen as in Figure 66, to show the delicate terminal regions of the branches of the enamel tubules. The prism decussation is more pronounced in this region. SEM: ×575.

uniform stripes running out to the surface at about 60° on the sides of the teeth (Figs. 74 and 75). On the labial surface, they were absent. Within each band, all the prisms appeared to run parallel, following an oblique course, with the prisms in adjacent bands inclined at roughly 90° to each other (Fig. 75). The SEM confirmed the regular shift in prism direction from band to band and strongly suggested that the course of the prisms was spiral (Figs. 73 and 75). This spiral would be such that each prism, on its path from the amelodentinal junction to the surface, would cross from one band to the next, probably more than once. The Hunter-Schreger bands extended to the surface of the enamel, but the outer enamel had a complex appearance in polarized light, apparently as a result of reduced form birefringence com-

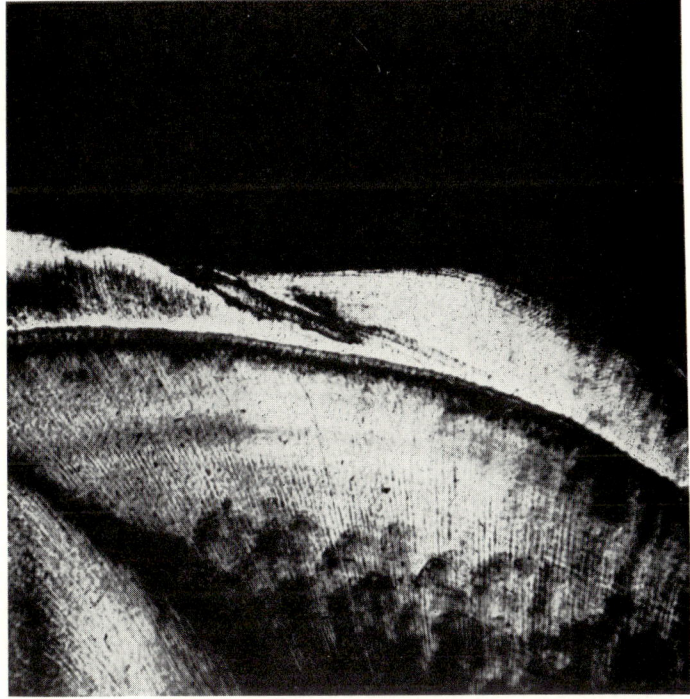

Figure 68. *Lemur*. Longitudinal ground section through lateral enamel and part of the coronal dentine. The enamel has the appearance of being made up of parallel, fairly straight prisms intersected by closely spaced incremental markings. In the middle of the field, a line runs obliquely through the enamel, which shows a surface furrow. Inside the line, there is a pronounced region showing positive birefringence. Below the mantle dentine layer, which is faintly birefringent, is seen the circumpulpal dentine, containing a number of rounded, arcade-shaped calcospherites. Crossed polars: ×110.

pared to the bulk of the enamel; this produced a reversal of sign (Fig. 74).

Cross-striations at 5 μm intervals were present along the whole length of each prism. In the innermost 20 μm of enamel, where the prisms were more or less at right angles to the amelo-dentinal junction, the cross-striations were in places so aligned as to give the effect of narrowly spaced incremental lines. Retzius lines were lacking. The enamel covers only the labial third of the tooth surface. At the margins, the enamel showed uniform,

254 *Evolutionary Changes to the Primate Skull and Dentition*

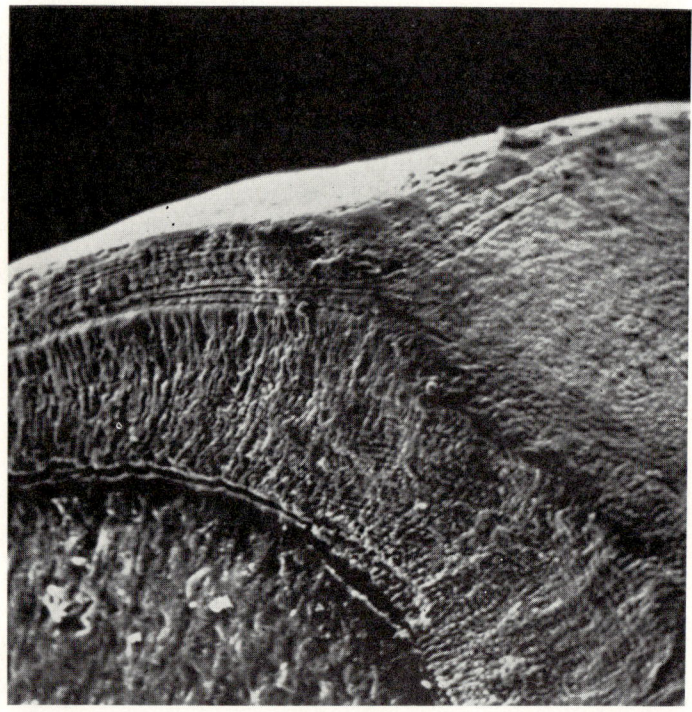

Figure 69. *Lemur*. Lower-power orientation view of the other half of the tooth figured in Figure 68. The oblique line through the enamel is seen at the left and runs into a facet ground tangentially in the enamel for the purpose of studying prism shape. The specimen has been acid-etched and reveals regular prism cross-sections on the tangential surface but prism bodies are obscure on the longitudinal face. Instead, this surface presents a pattern of rather wavy lines corresponding to the incremental markings. SEM: ×230.

strong negative birefringence relative to the surface normal (Fig. 76). The amelodentinal junction was almost smooth, showing scalloping only on a very small scale. In the plane of section employed, groups of tufts were observed at the origins of the Hunter-Schreger bands. Spindles were present but there were no enamel tubules.

Dentine

The mantle dentine was 10 to 15 μm thick in all species and was almost isotropic (Fig. 76). Except in *Galago* (see below), the hy-

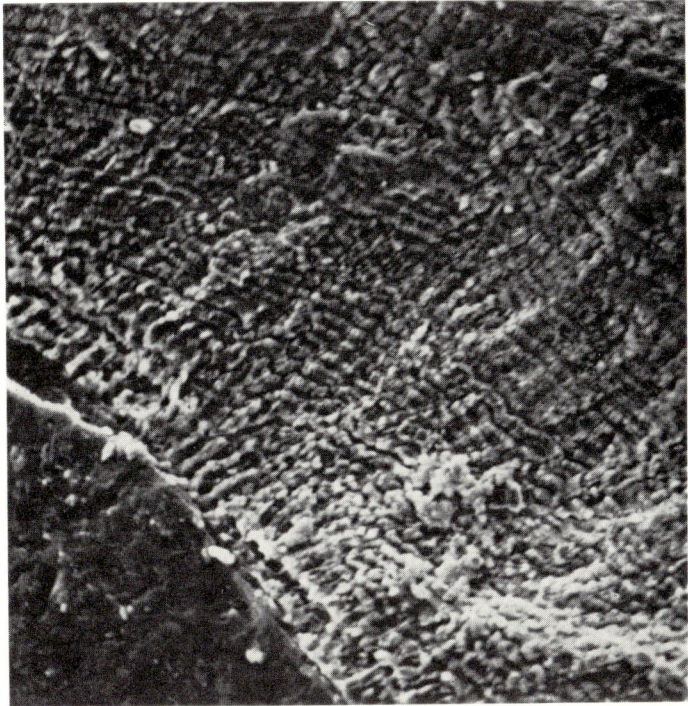

Figure 70. *Lemur*. Longitudinal surface below tangentially ground facet from Figure 69, to show the virtual absence of recognizable images of prisms and the dominant pattern of incremental markings. SEM: ×575.

aline layer was of a similar thickness but contained oblique coarse fibers running occlusally. A granular layer was present only in *Lemur* (Fig. 78), *Daubentonia* (Fig. 76) and *Propithecus* (in this species only in the apical region). These superficial layers contained small, elongated, hyperboloid calcospherites (Fig. 76). Such structures were not observed in the circumpulpal dentine of *Daubentonia*, which may be an effect of the plane of section, but were present in the inner two thirds of the coronal dentine of the other prosimians (Fig. 68). In this region, the calcospherites had the form of rounded arcades. Interglobular dentine was present in *Lemur*.

Von Ebner lines were difficult to distinguish in the coronal dentine of all species except for *Perodicticus* and *Propithecus*, where they were clear only near the cusps and were about 8 μm

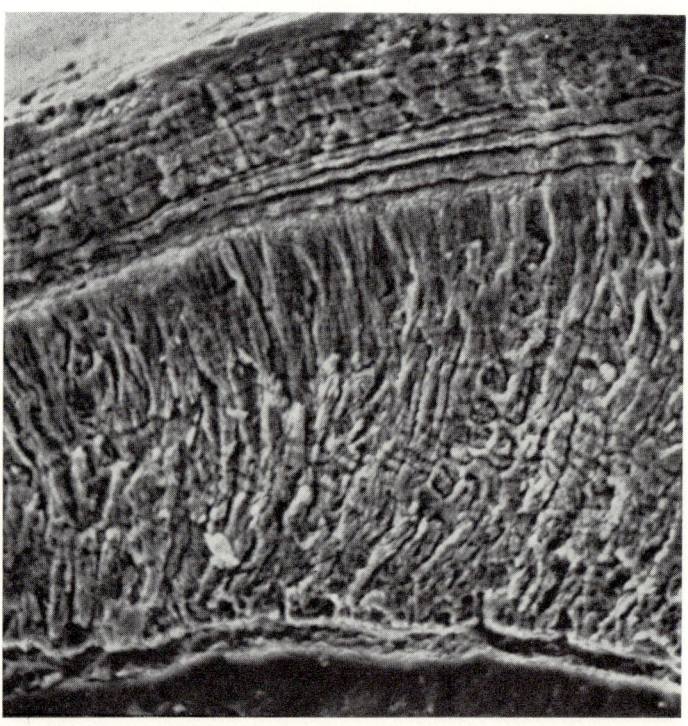

Figure 71. *Lemur*. Part of the specimen seen in Figure 69, to show the oblique line about two thirds of the way through the enamel from the amelodentinal junction. Inside the line, prism bodies are somewhat more obvious than in the region shown in Figure 70, but outside the line, the enamel appears to be prismless. Note the convergence of the inner incremental markings of the outer layer in the cervical direction (towards the left). SEM: ×575.

apart. In *Galago*, however, incremental lines were observed in the superficial root dentine. In this species, the hyaline layer was absent or very thin and the incremental lines ran out to the dentine surface. Here, they appeared to form perikymata-like steps, so that the dentine-cementum boundary had an undulating appearance (Fig. 77). Owen lines having the appearance described in *Saimiri* and the Callithricidae were present near the cusps in *Lemur, Galago,* and *Propithecus*.

Irregular secondary dentine was not observed in any of the prosimians. A layer of regular secondary dentine, on the other

Figure 72. *Lemur*. The prismless enamel outside the oblique line; closer view of the field seen in Figure 71. SEM: ×2,300.

hand, was present in both crown and root in all species, reaching a thickness of 20 to 30 μm. This tissue was not demarcated very clearly from the primary dentine. In *Daubentonia*, the outer third of the dentine contained straight, fine tubules. Inside this was a thicker, less birefringent layer in which the tubules became coarser towards the pulp. About halfway through this thick inner layer there was a sudden deviation of the tubules, through an angle of about 100°. In this species, therefore, the presence and extent of secondary dentine are open to question.

Cementum

Of the attachment tissues, only the cementum can be described, as all of the material consisted of teeth extracted from dried skulls; only fragments of periodontal ligament and bone were preserved (e.g. Fig. 77).

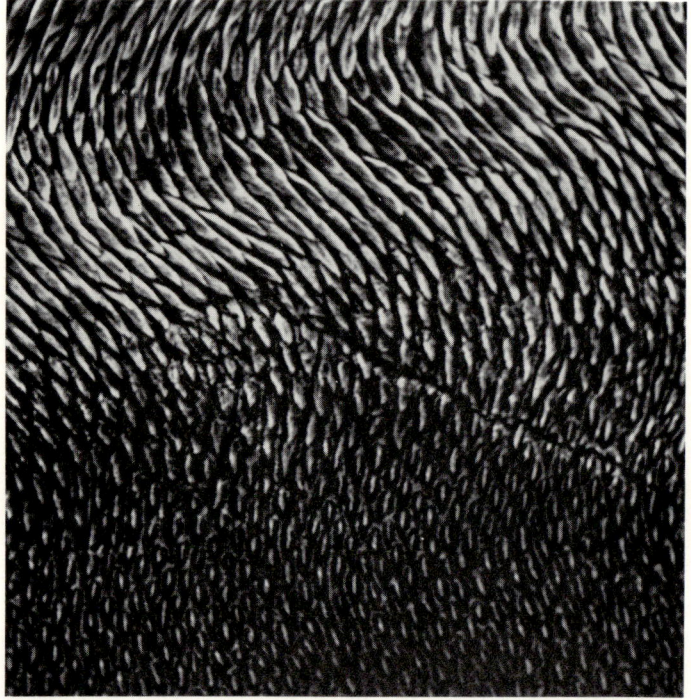

Figure 73. *Daubentonia*. Acid-etched surface ground tangentially to the enamel surface. Prisms are seen in oblique to transverse section at the junction between neighboring Hunter-Schreger bands. The prisms form, in this plane, sinuous rows, each separated by a wide layer of interprismatic enamel, which is strongly etched. SEM: ×550.

The cementum varied considerably in structure and in extent. The most regular form occurred in *Daubentonia*, where it constituted a layer about 15 μm thick over the whole surface of the dentine, overlapping the marginal enamel as well (Fig. 76). The birefringence was uniformly opposite in sign to the dentine, so that this layer would be recognized as primary cementum.

In *Lemur* and *Propithecus*, the cementum was more abundant, especially on the inner aspects of the roots, and reached a thickness of 50 to 60 μm although there was considerable variation in thickness (Fig. 77). The sign of birefringence was opposite to that of the dentine, indicating a predominantly radial fibre orientation, but the birefringence was extremely weak, so

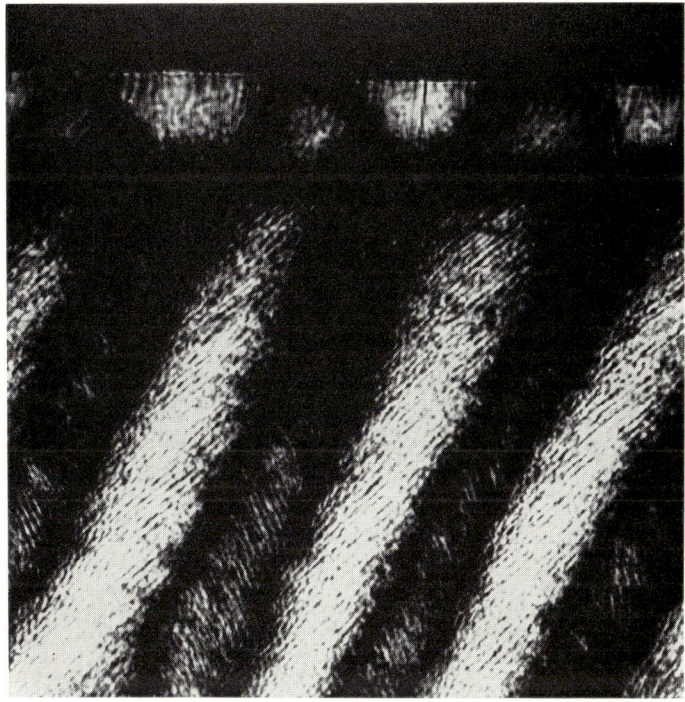

Figure 74. *Daubentonia.* Ground transverse section of incisor, showing the outer part of the enamel. The prisms are organized into extremely regular Hunter-Schreger bands. Note the alternating orientation of prisms in adjacent rows. The bands extend right to the surface but this is not completely clear because of the complex pattern set up in the superficial enamel by a reversal of birefringence. Crossed polars: ×280.

that the cementum fibres must form a fairly random meshwork. In *Lemur*, there were, however, well-marked Sharpey fibers in places and the cementum was stratified (Fig. 78). Cementocytes were few in *Lemur* (Fig. 78) and absent in *Propithecus*.

In *Galago* and *Perodicticus*, the cementum was almost isotropic and in the coronal parts of the roots, where it was very thin, was difficult to distinguish from the hyaline layer of the dentine (Fig. 77). The cementum layer was thicker on the inner aspects of the roots than on the outer aspects and also reached a greater thickness of about 50 μm around the apices (Fig. 77). Cementocytes were found only in these periapical masses.

260 *Evolutionary Changes to the Primate Skull and Dentition*

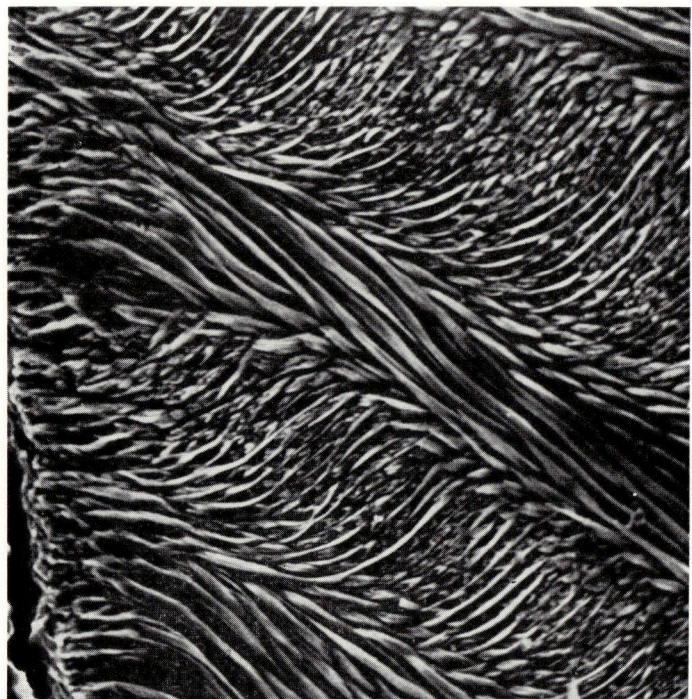

Figure 75. *Daubentonia*. Acid-etched ground surface transverse to axis of incisor, viz. perpendicular to the plane of Figure 73. Enamel, showing Hunter-Schreger bands, in which there is a strong suggestion of spiralling of the prisms. In this section plane, the interprismatic enamel is distinct in the diazones. Dentine is to the left. SEM: ×525.

DISCUSSION

In the existing literature, observations on primate dental tissues are limited and generally of a fragmentary nature. For instance, Boyde (1964) and Helmcke (1967) studied aspects of enamel structure in *Macaca* and Schmidt and Keil (1971) presented sundry observations on *Cebus* and *Cercopithecus*. Ockerse (1963) made a fairly detailed study of tooth structure in *Cercopithecus aethops*. Carter (1922) surveyed prosimian teeth, including a number of fossil specimens, but was concerned only with enamel structure, especially the occurrence and abundance of enamel tubules. Because they are valuable experimental animals for research into periodontal disease, descriptions of the

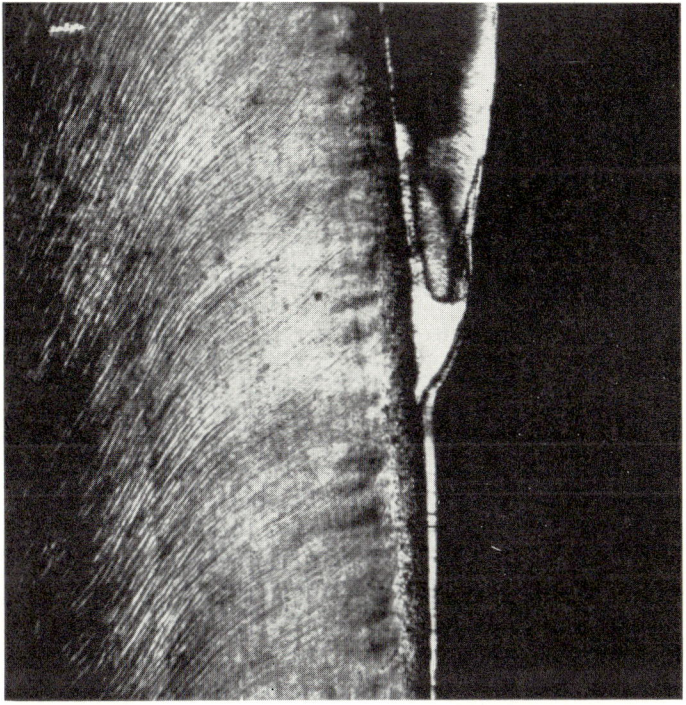

Figure 76. *Daubentonia*. Transverse ground section of incisor, showing the margin of the enamel, dentine, and cementum. The marginal enamel is strongly negatively birefringent and is overlapped by a layer of cementum, which is also uniform in appearance but of opposite sign. The superficial dentine is dark and comprises the mantle dentine, under the enamel, the hyaline layer under the cementum and, below the hyaline layer in turn, the granular layer. Hyperboloid calcospherites are present just inside the superficial dentine but the circumpulpal dentine largely lacks these structures. Crossed polars: ×110.

periodontal ligament and associated supporting tissues have been made for a number of species: *Alouatta caraya* (Hall, Grupe and Claycomb, 1967), *Cercopithecus* sp. (Smukler and Breyer, 1969), *Galago senegalensis* and *G. crassicaudatus* (Grant, Chase, and Bernick, 1973), *Macaca mulatta* (Arnim and Hagerman, 1953; Kindlova, 1965; Cutwright, 1970; Glenwright, 1970), *Papio anubis* (Avery & Simpson, 1973), *Callithrix jacchus* and *Saguinus fuscicollis* (Cohn, 1972; Levy, Dreizen and Bernick, 1972).

Thus, it seemed timely to carry out the survey described here,

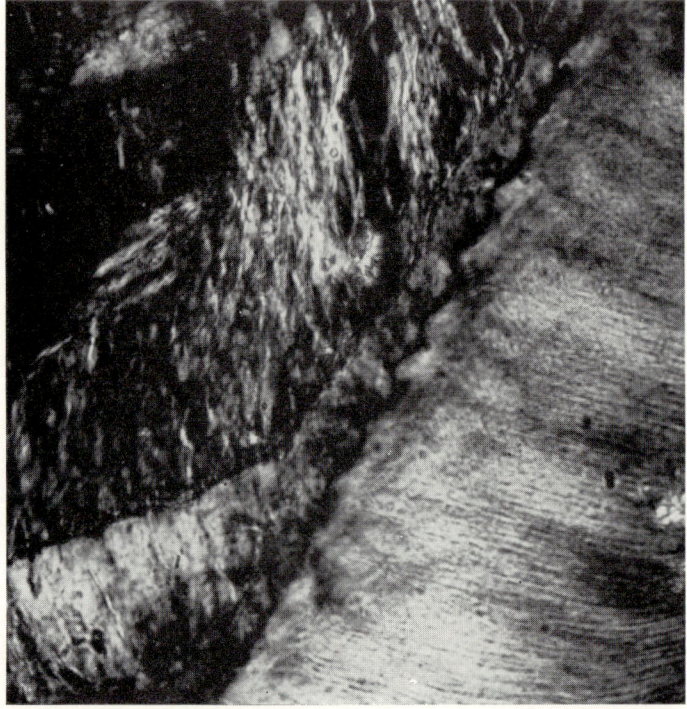

Figure 77. *Galago*. Longitudinal ground section through molar root, near the apex. In the superficial dentine, there are incremental markings which seem to reach the surface, giving the junction with the cementum a stepped appearance. The cementum is weakly birefringent and contains few cells. At this level, the thin layer found over most or the surface thickens rapidly into a mass which surrounds the apex (below). A remnant of the periodontal ligament, made up of wavy oblique fibers, has been retained, together with a fragment of alveolar bone (left). Crossed polars: ×280.

the aims of which were threefold. First, to discover whether primate dental tissues varied in such a way as to be of taxonomic value. Secondly, to increase knowledge of the phylogeny and adaptive radiation of the various tissues. Thirdly, to try to achieve a better understanding of the structure and development of each tissue, particularly with regard to function.

However, despite the amount of material examined, the present survey must be regarded as far from complete, in that some families were not included at all and, of those which were

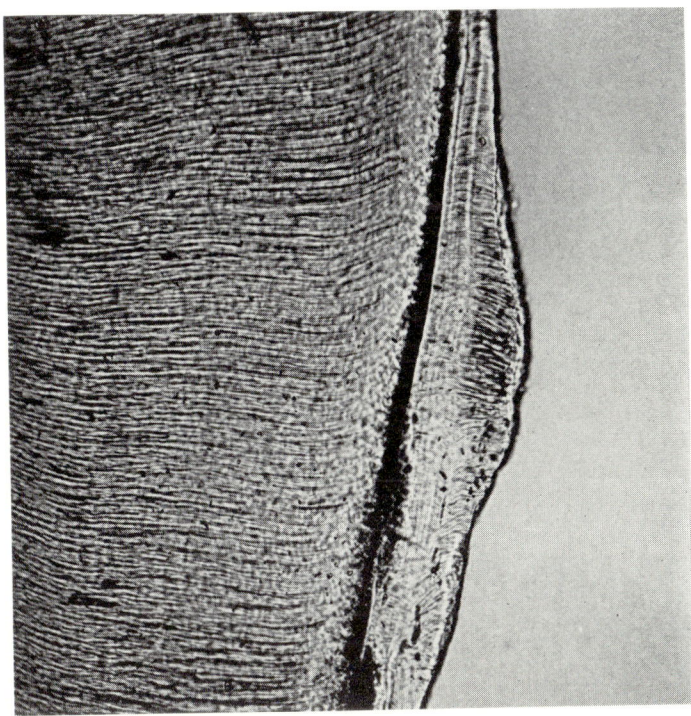

Figure 78. *Lemur*. Longitudinal ground section through molar root, showing dentine and cementum. The cementum is finely striated and thickens in places to form protuberances in which Sharpey fibers are numerous and in which a few cells are present. Note the gradual increase in prominence of the granular layer in the outer dentine towards the apical region (below). Bright field: ×280.

studied, most were represented by single specimens of only one or two species. Important intrafamilial and intraspecific variations must have been missed and thus it is perhaps not surprising to find that we can only make limited and tentative suggestions as to the potential taxonomic value of variations in dental tissues.

Considering our results as a whole, it appears possible to separate two groups of primate families on the basis of tooth structure. Group A would comprise the Hominidae, Pongidae, and Cercopithecidae; the Group B, the Callithricidae and Prosimii. The Cebidae show mixed characters of these two groups, the features of which are summarized in Table XIX. Of

TABLE XIX
PRINCIPAL TAXONOMIC FEATURES OF TWO PRIMATE GROUPINGS

Tissue	Group A: *Hominidae/Pongidae/Cercopithecidae*	Group B: *Callithricidae/Prosimii*
Enamel	Prisms: consistently pattern 3	Prism outlines tending to be complete, with increase in interprismatic regions.
	Decussation of prisms, forming Hunter-Schreger bands	Decussation limited, Hunter-Schreger bands ill-defined or absent.
	Cross-striations about 5 μm apart + Retzius lines 15-30 μm apart	Cross-striations about 5 μm apart, incremental markings 5-10 μm apart.
	Enamel spindles	Enamel spindles + enamel tubules in some species.
Dentine	Numerous calcospherites	Fewer calcospherites
	von Ebner lines prominent	von Ebner lines tending to be obscure
	Owen lines rare	Owen lines fairly common
	Granular and hyaline layers consistently present and well-defined	Granular layer frequently absent, as is hyaline layer in some cases.
Cementum	Well-defined primary and secondary layers of classical structure	Primary and secondary layers often not distinguishable; cementum structure ill-defined and variable.

the Cebidae, *Ateles* seems to be allied with Group A and *Saimiri* with Group B. This could mean that some of the structural differences listed in Table II are related to tooth size rather than to true taxonomic differences, since *Ateles* is appreciably larger than *Saimiri* as, on the whole, are the animals in Group A compared to those in Group B. This possibility merits further study although, even within the limits of this survey, there is some size overlap between the two groups; the teeth of *Lemur* and *Propithecus* approach those in *Cercopithecus* in size.

There are, of course, a number of obvious exceptions to the division given in Table II but some of these anomalies appear to be so distinctive as to have their own specific taxonomic value. For instance, *Gorilla* is easily distinguishable from *Pan*, partly by the lack of prism decussation, as well as by the extreme regularity of prism shape and by the radial striping of the root dentine. Likewise, *Daubentonia* is differentiated from the other prosimians by the arrangement of prisms in rows and by the occurrence of Hunter-Schreger bands which are, moreover, so highly organized that the enamel is quite distinct from that of all other primates.

There are other species of primates which, while fitting into the scheme of Table XIX, possess apparently characteristic features of tooth structure, although with study of further material, these may prove to be less specific or to be deviations from normal structure. Examples of such features are the curious enamel structure of *Lemur* (Figs. 69 to 72) and the swellings of the tubules in the cervical dentine of *Saguinus* (Fig. 62).

Enamel tubules were identified with certainty only in *Propithecus, Lemur, Galago,* and *Saguinus*. This agrees with the observations of Carter (1922) who identified a few tubules in *Perodicticus* also. It appears from the work of Carter that the presence and extent of enamel tubules could be a valuable aid to identification using tooth tissues. He found that well-developed tubules were present in lemurids, indrisids, lorisids, and in *Tarsius*, although there were exceptions; there were fewer tubules in *Propithecus* than in *Indris*, and fewer in *Perodicticus* and *Nycticebus* than in other lorisids (not named). In agreement with our results, Carter found no tubules in *Daubentonia* (*Chiromys*).

It is greatly regretted that the author's survey did not include the tree shrews, especially because of the controversy surrounding their inclusion with the primates (Napier & Napier, 1973; Osman Hill, 1972). Apart from the report by Carter (1922) that the enamel of *Tapaia* lacks tubules, no details of tooth structure are available. There is clearly a case for the comparison of tooth tissues in the tree shrews, prosimians, elephant shrews and flying lemurs, all of which are believed to be near relatives (Osman Hill, 1972).

With regard to taxonomy, our conclusion is that tooth structure is of value. The dental tissues of certain species are sufficiently characteristic to be readily recognized, and a particular combination of characters may well allow an unknown specimen to be assigned to one of two groups within the Primates. Establishment of the full taxonomic potential of tooth structure must, however, await a truly comprehensive survey. Tooth structure would certainly be of great value as a tool in identification, when the material is too fragmentary for tooth morphology to be reliably described. Under these circumstances, it would be essential to be able to assign the material to the correct order before more detailed classification was attempted by the use of descriptions such as those given in the present survey. Among the mammals, there is considerable variation in the details of tooth structure (Schmidt and Keil, 1971), and Boyde (1971) has enumerated, with examples, features likely to form the basis of a scheme for identification of tooth fragments. At present, however, the available information has not been fully systematized and much more information is required.

In the remainder of this discussion, we will consider aspects of the structure of the enamel, dentine, and cementum which appear to throw some light on development, phylogeny, or adaptation to function.

From an examination of Table XIX, it can be seen that there are a number of trends in enamel structure. The principal features are: a tendency for prisms to become more closely packed and to adopt an open, pattern 3, cross-section instead of a closed outline, an increase in the extent of prism decussation, and a reduction in the extent of tubule penetration, so that in most anthropoids only spindles are found and enamel tubules are lacking. These features may be all related to the adaptation of enamel to its function.

Enamel is a hard, unyielding material, the function of which is to reduce the rate of wear of the tooth and enable the teeth to deal with resistant foodstuffs. Essentially, the enamel forms a relatively incompressible cap seated on the much more elastic dentine; the latter tissue takes up stress by reversible deformation and has a much higher breaking stress than the enamel (Tyldesley, 1959). In reptiles, the enamel is made up of crystals standing more or less parallel to each other and perpendicular to the surface. This structure probably results in mechanical anisotropy; that is, a tendency to fail more readily in planes parallel with the crystals than in others. This is probably not a serious disadvantage in these animals because the teeth are simple in shape, are subjected to stresses mainly in the vertical plane because of the hinge-like jaw action and are replaced frequently (Poole and Shellis, 1976). In mammals, however, it is more important that mechanical anisotropy is reduced because, in these animals, the teeth are more complex in form and are usually subjected to a wide variety of stresses by the multidirectional movements of the mandible. Moreover, tooth replacement is limited so that the functional life of each tooth is greatly extended.

Mechanical anisotropy is reduced by the varied orientation of the constituent crystals. This also increases the toughness of the material by introducing internal surfaces, such as the prism junctions, which limit fracture (Poole, 1956) by acting as "crack stoppers" (Gordon, 1974). Dispersion of crystal orientation is achieved in three ways. Within each prism, the orientation changes across the width. In addition, the mean crystal orientation within the prisms is different from that within the interprismatic enamel, where this is present. Finally, diversification is achieved by prism decussation. It appears that, in primates, there may have occurred a stage when the decussation mechanism supplanted that relying on the presence of interprismatic enamel. Decussation involves movement of ameloblasts relative to each other during formation of enamel. Boyde (1964) suggested that the cells "follow" the eccentric growth of the prisms, but it seems likely that, in the evolution of decussation, there were modifications affecting the cells; for instance, changes in the pattern of intercellular attachments in the ameloblast layer, which at least facilitated ameloblast movement. A report by Kallenbach (1973), that the distribution of organelles at the surface of Tomes process is asymmetrical, raises the possibility that factors intrinsic to the ameloblasts play a part in prism decussation.

The most elaborate system of prism decussation was observed in the incisor of *Daubentonia*. This is of considerable interest not only because it is so unlike other prosimian enamels, but because of the convergence with the rodents and lagomorphs. These also have continuously growing and erupting incisors covered on the labial surface with enamel in which prism decussation is pronounced. In common rodents, such as rats and mice, decussation has been interpreted as lateral, two-dimensional curvatures of prism rows, curvature in neighboring rows being in opposite senses (Boyde, 1964; Schmidt and Keil, 1971). It seems to us, however, that the prism course in *Daubentonia* is more complex, with prism sheets apparently following a spiral course. The spiral seems to be not cylindrical but sheared, so that each prism spirals obliquely through the enamel. In section, the interfaces between Hunter-Schreger bands are regions where the spiralling prisms project perpendicularly on to the plane of section. Such a prism course clearly entails extremely complicated movements of the ameloblasts. In the incisor enamel of a number of rodents and lagomorphs, there is pronounced undulation of the prisms, producing patterns reminiscent of those seen in our preparations of *Daubentonia* (Korvenkontio, 1934; Schmidt and Keil, 1971) and it may be that the enamel structure in each of these three different types can be explained in the same way; Boyde (1964) pointed out that in the incisor enamel of the rodent *Myocastor*, the prisms appeared not to remain within the territory of the same Hunter-Schreger band. The fact that decussation is such a constant, and prominent, feature of continuously growing, procumbent gnawing incisors suggests structural adaptation. It is worth noting that some degree of prism torsion can often be discerned in a variety of preparations viewed in the SEM, even in the impressively straight prisms of *Gorilla*.

As a footnote to the foregoing, the authors would like to point out that, although the prisms in Group B primates often had outlines which were closed or almost so, the prism morphology is not strictly pattern 1 but rather is a variant of pattern 3, described as pattern 3c by Boyde (1964), in which the interprismatic regions are still continuous with the enamel of the prism bodies. This pattern is common in Carnivora (Boyde, 1964). Our own SEM observations on the hedgehog (*Erinaceus europaeus*), which has a true pattern 1 prism morphology, show that the acid etch pattern is quite distinct from that observed in prosimian enamel. Likewise, although the prisms in *Daubentonia* are arranged in rows (Figs. 52 and 54), they do not seem to be sufficiently closely packed within the rows for the prism structure in this species to be referred to pattern 2 (compare Fig. 52 with Fig. 4 of Boyde, 1971). Among the primates, therefore, the prism structure always seems to be a variant of pattern 3.

The nature and function of enamel tubules is unclear, despite a number of investigations. In addition to prosimians, enamel tubules are found in certain other placental mammals, e.g. hyraxes, in most marsupials, in a number of extinct therians and mammal-like reptiles (Moss, 1969) and in at least one living reptile, *Uromastyx* (Cooper and Poole, 1973). The view of Moss and Applebaum (1963), that the tubules are unmineralized rods of enamel, is untenable in view of the fact that they are branched and often cross the prism direction obliquely (Schmidt and Keil, 1971). Similarly, the same facts must cast doubt on the hypothesis of Lester (1970), that they arise as cytoplasmic extensions of the ameloblasts, left behind in the enamel as it is laid down, especially as the branches are directed outwards towards the enamel surface. The much earlier alternative hypothesis, more recently expounded by Schmidt and Keil (1971) and Osborne (1974), is that odontoblast processes extend into the enamel before mineralization is complete, and this hypothesis seems to us the more likely. It does, however, appear to conflict with the suggestion of Holland (1975) that, at least in cats, odontoblast processes begin to withdraw along the dentinal tubules soon after the first layer of dentine is formed and, thereafter, never reach more than a third to a half of the way along the tubules.

Our own observations on *Propithecus* (Figs. 65 to 67) confirm that the tubules are branched and show that they cross over from the dentine and that they contain organic material. The nature of the organic material is obscure; our specimen had been dried for several decades. Fosse and Holmbakken (1971) also observed organic tubule remnants in etched marsupial enamel from museum specimens. Osborne (1974) has suggested that the tubules evolved in animals with grinding dentitions and had a sensory function auxiliary to that of the periodontal ligament. There is no experimental evidence in support of this. It must also be borne in mind that, in grinding dentitions, enamel is rapidly worn away on the occlusal surface to expose the more vital dentine. Further, it seems anomalous that the only marsupial reported to lack enamel tubules, the wombat, has a grinding dentition. The only alternative suggestion to occur to us is that the function of the tubules is again mechanical. The presence of tubules in enamel could have two effects; (a) a reduction of the stiffness of the enamel; the results of Haines, Berry and Poole (1963) indicate that internal spaces of enamel might be responsible for a greater compressibility observed at low loads; (b) an increase in toughness by the "crack stopper" mechanism; it has been suggested that the tubules of dentine contribute in this way to the properties of the tissue (Renson, Boyde and Jones, 1974). In the prosimians, tubules were much more common in the occlusal and cuspal enamel. This could be related to the operation of different patterns of stresses in a newly erupted tooth, with intact occlusal enamel, and an older tooth, where the enamel essentially forms an annulus surrounding a concave dentinal surface. The reported absence of tubules in *Tupaia* (Carter, 1922) could indicate that enamel tubules are a secondary feature in prosimians, and this would be consistent with the sporadic distribution of tubules among the prosimians. This also conflicts with Osborne's assertion (Osborne, 1974) that the tubules were primitive structures lost during the evolution of placental mammals.

Incremental markings were an almost universal feature of the enamel in all the primates examined. Cross-striations were usually found along the length of the prisms. These markings seem to be the result of rhythmical variations in prism structure, such as crystal orientation or variation in pore structure, which renders them visible in polarized light and also leads to differential susceptibility to acid etching (see Figs. 3 and 38). Cross-striations have been related to a diurnal variation in ameloblast activity (Newman and Poole, 1974); the constancy of repeat distance at about 5 μm among primates suggests a uniformity of the rate of ameloblast activity in this group. In the primates of Group B, where the prisms were usually parallel because of the near absence of decussation, the cross-striations were generally aligned to produce the impression of oblique striae running through the enamel. This was also found in the outer enamel of Group A primates and in the inner enamel of *Daubentonia*, where the prisms did not decussate. In the animals of Group A there were, in addition, the more prominent striae of Retzius, spaced wider apart at 15 to 30 μm. The striae of Retzius probably result from greater disturbances of amelogenesis than those producing the cross-striations, possibly as a result of imperfect synchronization of two independent but interacting circadian rhythms (Newman and Poole, 1974). The variation in periodicity of the Retzius lines in the primates suggests that the frequency of the interaction responsible varies considerably, although some of the variation could be due to differences in plane of section. Retzius lines proper were not found in Group B primates but accentuation of some cross-striation lines were observed in some species, e.g. *Galago*, indicating that in these also some interaction between circadian rhythms occurs to affect the ameloblasts.

Dentine is phylogenetically the oldest of the dental hard tissues and its fundamental structure, as a feltwork of mineralized collagen fibers deposited parallel with the incremental lines, was established early in vertebrate evolution. Tissues such as osteodentine, plicidentine, and vasodentine are simple variations of dentine involving no profound modification of the tissue structure. The structural variety manifested in the dentine among primates is thus, not surprisingly, rather limited, but a number of points are of interest.

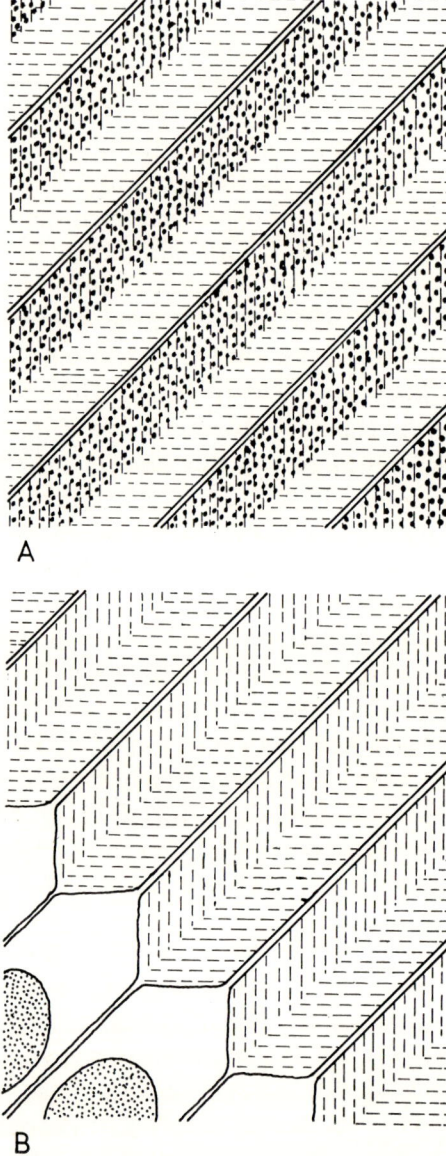

Figure 79. Diagrams to illustrate the structure of root dentine in *Gorilla*. (A) the suggested orientation of the fibers (broken lines, tubules are shown as parallel solid lines). The stippling indicates the extinction in polarized light of alternate stripes, in which the fibers are aligned parallel with the polarization plane of the polarizer prism, (B) the suggested mode of development of the structure, with the matrix fibers deposited parallel with the cell membrane surfaces of the odontoblasts.

In all the primates examined, the superficial coronal dentine, the mantle layer, was differentiated from the circumpulpal dentine and was relatively constant in thickness (15-30 μm), considering the variation in tooth size between the species. In all species, the layer tended to be weakly birefringent or isotropic, indicating that radial von Korff fibers were not the dominant constituents they are in certain other mammals, e.g. toothed whales (Boyde, 1971; Schmidt and Keil, 1971). A hyaline layer was present on the roots of most primates except for *Galago* (Fig. 77). Frequently, obliquely orientated fibers were distinguished in this layer. Similar fibers were illustrated in elephant tusk by Schmidt and Keil (1971). These authors drew no distinction between the mantle and hyaline layers and the author's results suggest that, in primates, there is usually little difference in their appearance under polarized light. However, recent observations suggest that, in man, the hyaline layer mineralizes centripetally (Owens, 1973; Kawasaki and Fearnhead, 1975), whereas the mantle dentine mineralizes centrifugally like the circumpulpal dentine (Kawasaki and Fearnhead, 1975). This phenomenon is possibly related to the presence of the obliquely oriented fibers in the hyaline layer.

The matrix fibers of circumpulpal dentine are laid down as a network parallel with the surface of the odontoblast layer (Lester and Boyde, 1967; Schmidt and Keil, 1971). In *Gorilla*, the root dentine showed striping parallel with the dentinal tubules under polarized light (Fig. 42). This structure may be the result of an unusually close parallelism between fibers and odontoblast surfaces. Our interpretation of the striped appearance is that it results from a coordinated kinking of the fibers in the longitudinal plane, as illustrated in Fig. 79a. This in turn could be due to matrix fibers being laid down strictly parallel with the conical distal surfaces of odontoblast cell bodies (Fig. 79b).

To some extent, the fibers and crystals of dentine are organized into laminae. The von Ebner lines represent alternating laminae, the fibers being parallel within each lamina but at an angle to those of the adjacent laminae (Schmidt and Keil, 1971). In the primates, the von Ebner lines were confined to the coronal dentine, only poorly defined fragmentary striations being found in the roots, except for *Galago* (see below). Further, the lines were noticeably less distinct in the animals of Group B than in those of Group A, indicating less regularity of dentine organization. However, the relative obscurity of the lines in Group B could also be due to their generally not being sectioned truly perpendicularly; the lines in apes and monkeys vary considerably in width and clarity between different regions of the tooth, presumably because of variation in plane of section. While von Ebner lines tend to be obscure in Group B primates, other lines, involving changes in fiber orientation over much greater distances, typically 30 or 40 μm, were quite common. Because these bands were always associated with synchronous deviations in the tubules (Fig. 56), they have been referred to as Owen lines. These lines are evidently not a reflection of incremental growth; they were never present over the whole crown and had a variable spacing. The fact that they were found only on the step flanks of the cusps of small molars suggests that they might result from a localized crowding of odontoblasts, with consequent disturbance of matrix deposition.

The parallel, regularly spaced lines in the peripheral root dentine of *Galago* had the appearance of incremental markings. In polarized light, they seemed to reach the surface of the dentine and form outcroppings comparable with perikymata on enamel surfaces (Fig. 77). This implies the complete absence of a hyaline layer and merits further study, but direct examination of the dentine surface by SEM, to verify the stepped appearance, would be hampered by the presence of the cementum layer.

The pattern of deposition of secondary dentine was somewhat variable. In most families, it was possible to detect a layer of regular secondary dentine in teeth which were moderately worn and had therefore been in function for some time. This ubiquity supports the generally held concept that the deposition of regular secondary dentine is a physiological process related to age. The distribution of irregular secondary dentine was much more sporadic and appeared to be confined to the primates in Group A. This tissue is often said to form rapidly in response to external stimuli, such as caries or attrition, reaching the pulp by way of the primary dentine. The irregular structure is held to be due to the rapidity of formation and the tissue is believed to seal off the affected region of dentine from the pulp. These concepts receive some support from the occurrence of irregular secondary dentine, in localized masses, in *Papio*, sealing off the exposed pulp chamber of the canines, and in *Macaca speciosa*, in the pulp horns underlying an attrition facet. However, many heavily worn teeth lacked irregular secondary dentine altogether and the tissue was present in some teeth where even the enamel was scarcely worn. In teeth where it was present, the preferred locations were the floor of the pulp chamber and the outer walls of the root canals. These results suggest that, while the concept of irregular secondary dentine as a reparative material holds good in some circumstances, formation of the tissue is often influenced by factors unknown at present. In this respect, irregular secondary dentine may have affinities with the various mineralized bodies formed within the pulp, denticles for example. Both tissues occur apparently randomly in some animals but consistently in others (e.g. *Cetacea;* Boyde, 1971; Schmidt and Keil, 1971).

The main trend in the structure of cementum among the primates appears to be a transition from a generally ill-defined, variable structure (Group B) to a more consistent organization (Group A) as a thin inner layer and a thicker outer layer, corresponding to the layers described as primary and secondary cementum in human teeth. The presence of cells does not seem to be a universally applicable criterion for distinguishing secondary from primary cementum. For instance, cells were almost completely absent from the cementum of *Gorilla*, while in *Pan*, cementocytes were very numerous and were found even in the innermost layers except in the cervical region. The abundance of cementocytes was also highly variable in the Cercopithecidae.

In man, the periodontal fibers are inserted into the cementum in two ways; as discrete bundles (Sharpey fibers) or as individual fibers but so closely packed that they make up the bulk of the cemental matrix (Selvig, 1965). The first mode predominates in secondary cementum, the second in primary cementum. In primates of Group B, much of the cementum was opposite in sign of birefringence to the dentine, indicating an overall orientation of the fibers perpendicular to the dentine surface, but the birefringence was very weak and Sharpey fibers were absent in many places. In this tissue, therefore, the periodontal fibers must be inserted individually, not as Sharpey fibers, but may be thin or dispersed, so that their birefringence is largely compensated by that of the so-called "intrinsic" fibers which run parallel with the tooth surface.

A consistent feature of primate cementum which may be of functional significance is that, where the cementum was thin, discrete Sharpey fibers were lacking so that the periodontal fibers were inserted individually; whereas in thicker layers, Sharpey fibers were prominent. This is strikingly illustrated by the eminences of the cementum layer in *Lemur* (Fig. 78) and, in other prosimians, by the appearance of Sharpey fibers only in the swollen masses of cementum surrounding the apices. In the Group A primates, the inserted fibers were closely packed in the cervical primary dentine but, apically, where it was overlain by secondary cementum, the primary layer showed an increasingly striped polarisation pattern, indicating a grouping of the attachment fibers into bundles (Fig. 44). Further, it usually became indistinguishable in appearance from the secondary cementum near the apex. It is not known whether there are differences in mechanical properties between a system of attachment in which the periodontal fibers are inserted essentially as individual fibers, and one in which they are grouped into bundles as Sharpey fibers. It may be significant, however, that in all the primates studied, cementum was consistently thicker, and Sharpey fibers therefore more prevalent, towards the apices and, to a lesser extent, on the inner aspects of molar roots.

It becomes clear that, apart from taxonomic implications, scrutiny of the dental tissues of primates highlights the fact that a whole series of problems remains unsolved. Of special interest and importance are those concerning the relationships between function and structure of a tissue, and between structure and its development. These particular problems apply to calcified tissues throughout the vertebrates. It is, therefore, the author's hope that this review will stimulate others to share an interest in trying to solve some of the mysteries.

REFERENCES

Arnim, S. S. and Hagerman, D. A.: The connective tissue fibres of the marginal gingiva. *J Am Dent Assoc, 47*:271-281, 1953.

Avery, B. E. and Simpson, D. M.: The baboon as a model system for the study of periodontal disease: Clinical and light microscopic observations. *J Periodontol, 44*:675-686, 1973.

Boyde, A.: *The structure and development of mammalian enamel.* Ph.D. Thesis, University of London, 1964.

Boyde, A.: The structure of developing mammalian dental enamel. In M. V. Stack and R. W. Fearnhead (Eds.): *Tooth Enamel.* Bristol, Wright, 1965, pp. 163-167, 192-194.

Boyde, A.: Comparative histology of mammalian teeth. In Dahlberg A. (Ed.): *Dental Morphology and Evolution.* Chicago, Chicago U Pr, 1971, pp. 81-94.

Boyde, A. and Lester, K.: The structure and development of marsupial enamel tubules. *Z Zellforsch, 82*:558-576, 1967.

Carter, J. T.: On the structure of enamel in the primates and some other mammals. *Proc Zool Soc Lond,* 599-608, 1922.

Cohn, S. A.: A re-examination of Sharpey's fibres in alveolar bone of the marmoset (*Saguinus fuscicollis*). *Archs Oral Biol, 17*:261-269, 1972.

Cooper, J. S. and Poole, D. F. G.: The dentition and dental tissues of the agamid lizard, *Uromastyx. J Zool Lond, 169*:85-100, 1973.

Cutwright, D. E.: The morphogenesis of the vacular supply to the permanent teeth of *Macaca rhesus. Oral Surg, 30*:284-291, 1970.

Fearnhead, R. W. and Stack, M. V.: *Tooth Enamel II.* Bristol, Wright, 1971.

Fosse, G. and Holmbakken, N.: Fibrils in marsupial enamel tubules. *Z Zellforsch, 115*:341-350, 1971.

Glenwright, H. D.: Observations on circular and longitudinal gingival collagen fibres in the rhesus monkey. *Dent Practic 20*:337-341, 1970.

Gordon, J. E.: *The New Science of Strong Materials.* Harmondsworth, Penguin, 1974.

Grant, D. A., Chase, J., and Bernick, S.: Biology of the periodontium in primates of the *Galago* species. *J Periodontol, 44*:540-550, 1973.

Haines, D. J., Berry, D. C., and Poole, D. F. G.: Behaviour of tooth enamel under load. *J Dent Res, 42*:885-888, 1963.

Hall, W. B., Grupe, H. E., and Claycomb, C. K.: The periodontium and periodontal pathology in the howler monkey. *Archs Oral Biol, 12*:359-365, 1967.

Helmcke, J. G.: Ultrastructure of enamel. In Miles, A. E. W. (Ed.): *Structural and Chemical Organization of Teeth.* London, Acad Pr, Vol. II, Chap. 15, pp. 135-163.

Hill, W. C. Osman: *Evolutionary Biology of the Primates.* New York, Acad Pr, 1972.

Holland, R.: *The dentinal tubules of the cat and their contents.* Ph.D. Thesis, University of Bristol, 1975.
Kallenbach, E.: Fine structure of Thomes' process of the cat ameloblast. *J Dent Res,* 54:Special Issue A, Abstr. L63, 1975.
Kawasaki, K. and Fearnhead, R. W.: On the relationship between tetracycline and the incremental lines in dentine. *J Anat,* 119:49-59, 1975.
Kindlová, M. The blood supply of the marginal periodontium in *Macaca rhesus. Archs Oral Biol,* 10:869-874, 1965.
Korvenkontio, V. A.: Mikroskopische Untersuchungen an Nagerincisiveren unter Hinweis auf die Schmelzstruktur der Backenzähne. *Ann (bot-zool) Soc Zool-bot Fenn Vanamo* 2:1-274, 1934.
Lester, K. S.: On the nature of 'fibrils' and tubules in developing enamel of the opossum, *Didelphis marsupialis. J Ultrastruct Res,* 30:64-77, 1970.
Lester, K. S. and Boyde, A.: Electron microscopy of predential surfaces. *Calc Tiss Res,* 1:44-54, 1967.
Levy, B. M., Dreizen, S. and Bernick, S.: *The marmoset periodontium in health and disease.* In Myers, H. M. (Ed.): *Monographs in Oral Science.* Basel, S. Karger, Vol. 1, pp. 6-45.
Melcher, A. H. and Bowen, W. H.: *Biology of the Periodontium.* New York, Acad Pr, 1969.
Miles, A. E. W.: *The Structural and Chemical Organisation of Teeth.* Volumes I and II. New York, Acad Pres, 1967.
Moss, M. L.: Evolution of mamalian dental enamel. *Am Mus Novit,* No. 2360, 1-39, 1969.
Moss, M. L. and Applebaum, E.: The fibrillar matrix of marsupial enamel. *Acta Anat,* 53:289-297, 1963.
Napier, J. R. and Napier, P. H.: *A handbook of Living Primates.* New York, Acad. Pr, 1973.
Newman, H. N. and Poole, D. F. G.: Observations with scanning and transmission electron microscopy on the structure of human surface enamel. *Archs Oral Biol,* 19:1135-43, 1974.
Ockerse, T.: The histology of the teeth of the vervet monkey. *J Dent Assoc S Afr,* 18:1-6 (suppl.), 1963.
Osborn, J. W.: Tooth sensitivity. *J Dent Res,* 53:1063, 1974.
Owens, P. D. A.: Mineralization in the roots of human deciduous teeth demonstrated by tetracycline labelling. *Archs Oral Biol,* 18:889-897, 1973.
Poole, D. F. G.: The structure of the teeth of some mammal-like reptiles. *Quart J Micr Sci,* 97:303-312, 1956.
Poole, D. F. G.: The use of the microscope in dental research. *Brit Dent J,* 121:71-79, 1966.
Poole, D. F. G.: An introduction to the phylogeny of calcified tissues. In Dahlberg, A. (Ed.): *Dental Morphology and Evolution.* Chicago, Chicago, U Pr, 1971, pp. 65-79.

Poole, D. F. G. and Shellis, R. P.: Eruptive tooth movements in non-mammalian vertebrates. In Poole, D. F. G., and Stack, M. V. (Eds.): *The Eruption and Occlusion of Teeth*, London, Butterworths, 1976, pp. 65-80.

Renson, C. E., Boyde, A., and Jones, S. J.: Scanning electron microscopy of human dentine specimens fractured in bend and torsion tests. *Arch Oral Biol*, 19:447-454, 1974.

Schmidt, W. T. and Keil, A.: *Polarizing Microscopy of Dental Tissues.* Oxford, Pergamon, 1971.

Scott, J. H. and Symons, N. B. B.: *Introduction to Dental Anatomy.* Edinburgh, Livingstone, 1974.

Selvig, K. A.: The fine structure of human cementum. *Acta Odont Scand.*, 23:423-441, 1965.

Smukler, H. and Dreyer, C. J.: Principal fibres of the periodontium. *J Periodont Res*, 4:19-25, 1969.

Symons, N. B. B. (Ed.): *Dentine and Pulp: Their Structure and Reaction.* Edinburgh, Livingstone, 1968.

Tydesley, W. R.: The mechanical properties of human enamel and dentine. *Br Dent J* 106:269-278, 1959.

INDEX

A

Adapidae, 6, 43, 94, 146
Aegyptopithecus, 49
Age, 28
Agenesis, 176
Alligators, 4
Allometry, 29, 159
Alouatta, 58, 71, 102, 141, 261
Alveolar bone, 216
Alveolar prognathism, 55, 91
Ameloblasts, 199, 267, 269, 271
Amelodentinal junction, 198
Anathana, 42
Anodontia, 176
Anthropoidea, 5, 32, 44, 94
Apatite crystals, 198, 204
Arcadal index, 116
Archaeolemur, 43, 94
Area dental arch, 117
Ateles, 238, 265
Attrition, 106, 150, 170, 172, 243, 274
Australian aborigine, 50
Australopithecus, 20, 24, 25, 26, 33, 49, 51, 55, 59, 62, 63, 64, 65, 66, 89, 103, 120, 123, 141, 142, 144, 149, 160, 161, 167, 168

B

Bands—Hunter Schreger, 199
Bands—Retzius, 221
Basic rectangle of mandible, 119
Bite force, 88
Brachiation, 60
Brachycephalization, 51, 61
Brain growth, 10
Brain size, 3-34
  Age changes, 10-11
  Cerebral Rubicon, 11-15
  Cranial capacity, 18-23
  Evolutionary changes, 4-8
  Metric definition, 8-10
  Racial differences, 27-30
Brain size and intelligence, 28
Brain weight, 5, 9
Brain weight and body weight ratio, 16, 34
Broca area, 18
Browridge, 90

C

Caenopithecus, 94
Calcospheriteas, 214, 224
Callithicidae, 141, 239, 261, 263
Canines, 140, 142, 161
Carabelli trait, 146-149
Carnivora, 269
Catarrhines, 5, 7, 9, 25, 44, 54, 98, 237
Cebidae, 9, 260, 265
Cellular cementum, 217
Cementoblasts, 216
Cementum, 197, 217, 227, 235, 236, 239, 243, 257-259, 275
Cephalix index, 28, 50
Cephalometric analysis, 70
Cercopithecidae, 44, 69, 73, 182, 260, 261, 263, 265
Cerebral Rubicon, 11-18
Cheek teeth, 142-149
Chin, 99-101
Circumpupal dentine, 213
Coefficients of racial likeness, 157
  of variation, 156
Cold adaptation, 50
Colobinae, 44, 182
Correlations
  Body height and skull shape, 52
  Brain size and stature, 27
  Brain weight and body weight, 16, 34
  Cranial capacity and motor development, 19
  Skull, 74-77

[281]

Cranial
  Balance, 56-61
  Base, 41
  Capacity, 18-23
  Correlation, 74-77
  Length, 63
  Size, 42-50
  Variation, 50, 70-74
Cranio-facial growth, 73
Crests, 43, 53, 65-67
Crown measuration, 150-159
Culture, 23-27

D

Daubentonia, 30, 140, 151, 226, 255, 257, 258, 265, 266, 268, 271
Deciduous teeth, 198
Deflecting wrinkle, 145
Dental arch, 75, 107-124
  Arch index, 114
  Area, 117
Dental evolution, 159-170
Dental histology, 219-260
Dental papilla, 210
Dental tissues
  Anthropoidea, 219-228
  Callitricidae, 239-244
  Cebidae, 237-244
  Cercopithecoidea, 228-237
  Prosimii, 244-260
Dental tubule, 211
Dental variability, 152-159, 162-165
Dentine, 210, 213, 219-260, 271-274
Dentition, 49, 92, 93, 130, 184
  Attrition, 170
  Development, 130-138
  Eruption sequence, 173-175
  Evolution, 159-170
  Odontometry, 150-159
  Tooth number, 175-184
  Tooth shape, 148-149
Diastema, 107, 123
Diazones, 199
Diet, 97, 171
Discriminant analysis, 69, 97, 134, 157
Dolicephalization, 51, 61
Dryopithecus, 49, 91, 92, 119, 143

E

Enamel, 197, 198, 210, 219, 222, 228, 232, 239-241, 244-254, 266-271
  Prisms, 198, 205, 220, 237
  Rods, 198
  Spindles, 201
  Striation, 199
  Tubules, 240, 265
Endocasts, 9
Endocranial capacity, 5
Endocranium, 3
Eocene, 43, 94
Eruption sequence, 173-175
Eskimo, 65
Eutherian dentition, 130, 138
Evolution of brain, 4-8
Evolutionary rates, 165-170
External Pterygoid muscle, 99, 100
Extra neurones, 20, 21

F

Face, 50
Face width, 76
Facial asymmetry, 72
  Prognathism, 76, 102
  Variation, 72
Factor analysis, 69
Field theory, 132
Foramen magnum, 41, 61-62
Functional matrix, 96

G

Galago, 7, 244, 254, 256, 261, 265, 271, 273, 274
Gallithrix, 7
Genetic control of tooth form, 131-136
Genetic factors, 70
Gigantopithecus, 77, 142
Glenoid fossa, 104
Gorilla, 7, 11, 13, 15, 20, 24, 31, 41, 46, 49, 53, 56, 58, 60, 62, 63, 66, 72, 73, 89, 91, 98, 106, 114, 117, 122, 147, 151, 175, 182, 219, 220, 221, 222, 225, 226, 265, 273, 275
Granular layer of Tomes, 214, 224

H

Hardy-Weinberg law, 134

Head balance, 56-61
Head form, 85
Heidelberg mandible, 98
Hominidae, 49, 53, 62, 263
Hominization of brain, 15
Hominoidea, 138, 163
Homo erectus, 10, 13, 14, 25, 26, 33, 49, 51, 59, 61, 63, 64, 65, 119, 139, 141, 142, 144, 147, 149, 160, 161, 166, 167
Homo neanderthalensis, 63
Homo rhodesiensis, 63
Homo sapiens, 3, 7, 15, 16, 20, 25, 26, 31, 32, 49, 50, 60, 87, 91, 119, 132, 139, 142, 151, 152, 161, 162, 166, 167, 173, 174, 176, 179, 183
Hunter Schreger bands, 220, 228, 239, 251, 254, 263, 265
Hyaline-layer of Hopewell-Smith, 213
Hydroxyapatite, 198-204
Hylobatinae, 10, 12, 46, 53, 114, 117

I

Incisors, 99, 138-140, 160-161
Incisor shovelling, 133, 139
Incisor tapering, 139
Indices:
  Arcadial 116
  Body weight and skull shapes, 52
  Brain size and stature, 27
  Brain weight and body weight, 16, 34
  Cephalic, 28, 50
  Cranial, 74-77
  Cranial capacity and motor development, 19
  Dental arch, 117
  Palatal, 113
Indriidae, 246, 265
Insectivores, 3, 5
Intelligence quotient, 28
Interglobular dentine, 214
Interproximal attrition, 172

J

Jaw recession, 55

K

Kenyapithecus, 119

Klinorhynchy, 89
Knuckle-walking, 60

L

La Chapelle-aux-Saints, 34
Language, 25
Lemur, 6, 43, 54, 55, 103, 248, 255, 256, 258, 259, 265, 276
Le Moustier skull, 109
Line-neonatal, 199
Loris, 30, 43

M

Macaca, 15, 63, 71, 73, 89, 93, 122, 136, 152, 175, 228, 260, 261, 275
Malocclusion, 109
Mandible, 94-98
Mandible-Basic rectangle, 119
Mandibular angle, 98-99
Mandibular condyle, 87, 101, 105
Mandibular growth, 95
Mantle dentine, 212, 233, 238
Masseter Muscle, 98, 100
Masticatory system, 85-124
  Chin, 99-101
  Dental arch, 107-124
  Mandible, 94-98
  Mandibular angle, 98-99
  Maxilla, 90-94
  Occlusofacial relationships, 102-103
  Stress, 85
  Temporomandibular joint, 103-107
Mastoid process, 62
Maxilla, 90-94
Maxillary-alveolar index, 112
Maxillary arch, 92
Maxillary growth, 92
Megaladapis, 30, 43
Merism, 131
Mesocephalization, 51
Microcebus, 6
Miocene, 8, 24, 49, 91
Mio-Pliocene, 24
Molar form, 143
Molar teeth, 101, 161
Molarization, 142
Multivariate analysis, 50, 68, 69, 97, 134, 157
Myocastor, 268

N

Nasal form, 54-56
Neanderthal, 28, 60, 63
Necrolemur, 7
Negro, 27
Neocortex, 4, 40
Neurocranium, 3, 40
New World monkeys, 7, 9, 44, 55
Nuchal crest, 59
Nycticebus, 265

O

Occlusal attrition, 170
Occlusion, 103
Occlusofacial relationships, 102-103
Occipital condyles, 42, 56, 59, 61, 71
Odontoblasts, 210, 211, 212, 269
Odontometry, 150-159
Old World monkeys, 5, 7, 9, 25, 44, 54, 98, 237
Olfaction, 6, 30
Oligocene, 32, 49
Oligodontia, 176
Omomyidae, 140
Oppossum, 4
Optic axis, 53
Orthodentine, 197, 271
Orthograde posture, 41
Owen's lines, 215

P

Palaeopropithecus, 30
Palatal index, 113
Palatal vault, 103
Palate, 50
Paleocene, 98
Pan, 11, 12, 13, 15, 20, 31, 41, 46, 48, 49, 53, 60, 63, 64, 66, 73, 89, 91, 98, 117, 119, 122, 151, 175, 220, 221, 222, 226, 265, 270
Papio, 10, 55, 63, 114, 182, 228, 229, 231, 232, 234, 261, 275
Paramolar tubercle, 149
Paranthropus, 167
Parazones, 199
Perikymata, 200
Periodontal ligament, 197, 216, 217, 228, 231, 232, 237, 239, 273
Peritubular dentine, 216
Perodictus, 6, 244, 246, 255, 258, 259, 265
Platyrrhines, 7, 9, 44, 55
Pleistocene, 24, 43, 49, 86
Plesiadapis, 6, 43
Plicidentine, 271
Pliocene, 24
Polarizing microscope, 209
Pongidae, 5, 29, 49, 60, 62, 87, 219, 263
Pongo, 11, 12, 31, 53, 56, 58, 66, 71, 91, 92, 117, 147, 151, 182
Posture, 41
Predentine, 211
Premolars, 161
Premolar analogy theory, 131
Primary dentine, 215
Prism, 199, 203, 205, 207
Progalago, 6, 43
Prognathism, 76, 102
Propliopithecus, 140
Propithecus, 255, 258, 265, 270
Prosimii, 263
Protostylids, 149
Ptilocereus, 6

R

Race, 27-30
Ramapithecus, 91
Ratios
  Brain weight and body weight, 16, 34
  Body height and skull shape, 52
  Brain size and stature, 27
  Cranial capacity and motor development, 19
Reptiles, 199, 267
Retzius lines, 199, 206, 221, 232, 271
Rhinarium, 54

S

Sagittal crests, 43
Saguinus, 239, 261
Saimiri, 58, 71, 256, 265
Scanning electron microscopy, 218
Secondary dentine, 215, 256
Sexual dimorphism, 135, 141